Virgil Coblentz

The Newer Remedies

A Reference Manual for Physicians, Pharmacists and Students

Virgil Coblentz

The Newer Remedies
A Reference Manual for Physicians, Pharmacists and Students

ISBN/EAN: 9783742818706

Manufactured in Europe, USA, Canada, Australia, Japa

Cover: Foto ©berggeist007 / pixelio.de

Manufactured and distributed by brebook publishing software
(www.brebook.com)

Virgil Coblentz

The Newer Remedies

A REFERENCE MANUAL

FOR

PHYSICIANS, PHARMACISTS & STUDENTS,

BY

VIRGIL COBLENTZ, A.M., Phil. D., F. C. S., etc.,

*Professor of Pharmacy and Pharmaceutical Chemistry in the New York College
of Pharmacy; Author of Handbook of Pharmacy; Member of the
Chemical Societies of Berlin and London; Fellow of the
Society of Chemical Industry, etc., etc.*

SECOND EDITION,
Revised and Enlarged.

NEW YORK:
D. O. HAYNES & CO.
1896.

PREFACE TO SECOND EDITION.

During the past year a large number of new remedies have been introduced, and, by reason of new data concerning many older ones of which our knowledge was but fragmentary, a new issue of this book has become a necessity. With but few exceptions, the author has omitted all articles which are to be found in the dispensatories and like standard books of reference.

NEW YORK CITY, APRIL, 1896.

PREFACE TO FIRST EDITION.

The constantly increasing number and complexity of new remedies, with the scattered condition of the literature on the subject, render it an almost impossible task for anyone except the specialist to become at all acquainted with them. The physician who keeps in progress with the advance of the times is constantly on the alert for new and improved remedies with which to combat disease. It is, therefore, the duty of the apothecary to be conversant with these, not only by title, but as to identity, sources, properties and doses. It has been the endeavor of the author to present as complete a list of these newer remedies as possible in a concise and alphabetical form. The various articles are arranged under their commercial names, including, however, in this arrangement all the more important synonyms.

In all instances where possible or practical, the sources and methods of preparation have been given, followed by such tests of identity as melting and boiling points, with solubilities, incompatables, medicinal properties and doses as far as known. Details concerning physiological action and clinical data could not satisfactorily be included in a work of this size, hence the physician, when full accounts are desirable, must necessarily be referred to current medical literature or more extensive works on modern materia medica. It should be observed that these newer remedies have appeared and continue to appear in rapid succession, and in many instances the most important data are wanting, some going no further than mentioning title and nature of the substance, others stating medicinal uses only, indicating that they are still in the experimental stage. Hence in view of this, it will be necessary to issue revised editions from time to time as the fund of information increases

The doses quoted are the minimum and maximum for an average adult, being lessened accordingly for children and the feeble, and for hypodermic use. The sizes of the doses are expressed in the metric system, including, however, their nearest equivalents (in round numbers) in apothecaries' weight.

Many of these remedies now offered are of a proprietary nature, hence all that is known of their composition is what is given upon the authority of various published analyses.

The author is indebted to Geo. C. Diekman, M. D., for his careful perusal of that portion relating to the medicinal properties and doses of the various remedies

NEW YORK CITY, APRIL, 1895.

THE NEWER REMEDIES.

ABRASTOL. (ASAPROL) (C_{10} H_6 OH. SO_2) $_2$Ca.

The calcium salt of beta-naphthol-sulphonic acid. It is a soluble white powder, and because of its antiseptic properties, it is employed as a preservative agent for foods. It is also employed as an intestinal antiseptic in doses of 1 to 2 Gm. (15-30 grains).

ABRIN. (JEQUIRITIN).

Abrin is the most **potent** principle contained in the Jequirity seed (*Abrus precatorius*). It partakes of the nature of an albuminoid and is very poisonous. Forms a brownish yellow soluble powder; employed to limited extent for producing artificial conjunctivitis. Fatal dose is 1-100 grain.

ACETAL. CH_2-CH(OC_2 H_5)$_2$.

Synonyms: Ethyliden-di-ethyl-ether; Di-ethyl acetal.

This body is obtained by the reaction between acetic-aldehyde and alcohol, whereby water separates. It is a colorless liquid, boiling at 104°-106° C. (219°-222° F). Soluble in 18 parts of water (25° C.) and very soluble in alcohol. It is employed as sedative and hypnotic. Dose is from 5 to 10 Gm. (77-154 grains).

Usually administered in form of an emulsion.

ACETAMIDO-ANTIPYRINE.

By the action of nitric acid antipyrine is converted into nitro-antipyrine; this is reduced by means of zinc and acetic acid to amido-antipyrine. This latter compound, when heated with sodium acetate and acetic anhydride, is converted into acet-amido-antipyrine, which forms yellow crystals, melting at 100° C. (222° F.) It is soluble in water and alcohol, and is recommended as an antipyretic in the same doses as antipyrine.

ACETAMINOL. C_6 H_3 (OCH_3) (C_3 H_5) O-CO. $C_6H_4NHCOCH_3$.

Synonym: p-Acetamidobenzoyl-Eugenol.

By interaction between p-nitrobenzoylchlorid and eugenol-sodium in molecular proportions, p-nitrobenzoyleugenol is formed, this upon reduction yields the corresponding p-Amidobenzoyl-eugenol, which is acetylated by means of acetic anhydride. This compound appears in the form of white scales or a crystalline powder, of melting point 100° C. It is almost insoluble in water, quite soluble in alcohol.

It is employed in treatment of phthisis.

ACETANILID. C_6H_5 NH.COCH$_3$.

Synonyms: Antifebrine; **Phenylacetamide.**

Obtained by prolonged interaction **between** pure aniline and glacial acetic acid at boiling temperature. Twenty grammes of aniline are boiled with 30 grammes of glacial acetic acid under an inverted condenser for from six to ten hours, till a sample of the mixture when removed solidifies on cooling, to a crystalline mass. The fused mass is poured into cold water, and the crystals which separate are filtered off and recrystallized from hot water or alcohol. When pure, acetanilid forms lustrous rhombic tables without odor or color, melting at 113° C. (235.4° F.), soluble at 15° C. (59° F.), in 194 parts of water, and in 5 parts of alcohol; in 18 parts of boiling water, and in 0.4 parts of boiling alcohol. Acetanilid should not be left in contact with spirits of nitre any great length of time ; with antipyrine it forms a pasty mass ; in aqueous solution with the alkali bromides and iodides it forms insoluble compounds.

Its properties are those of an antipyretic. The average dose is from 0.2-0.5 Gm. (3 to 8 grains). The various derivatives of acetanilid employed in medicine are Asepsin, Iodantifebrin, Antinervin, Benzanilid, Exalgin.

Among the various remedies which are supposed to contain acetanilid as one of their constituents are Antikamnia, Phenolid, Exodyne, Antikol, Pyretin, Phenatol, Kaputin, etc.

ACETON.

A proprietary **grip and headache remedy. Not to be confounded with acetone.**

ACETONE. $CH_3-CO-CH_3$.

Synonym: Di-methyl-ketone.

This is prepared by the dry distillation of calcium acetate; it boils at 56° C. (132.8° F.), has a peculiar ethereal odor and sharp burning taste. Miscible with water, alcohol and ether.

It is employed as a nervine in doses of 5 to 15 minims, in water or infusion of valerian.

ACETONORESORCIN. $C_{15}H_{16}O_4+H_2O$.

A combination of two molecules of resorcin and one of acetone. Used as an antiseptic. Small anhydrous prisms, soluble in alkalies, insoluble in water and alcohol; melts at 212° C. (413.6° F.)

ACETOPHENONE. See Hypnone.

ACET-ORTHO-TOLUID. $C_6H_4 (CH_3) NH COCH_3$.

Synonym: Ortho-Tolyl-acetamid.

This is an isomeride of exalgine. obtained by prolonged interaction between ortho-toluidin and glacial acetic acid at boiling temperature. This occurs in colorless needles; melting point, 107° C. (224.6° F.) Soluble in hot water, alcohol and ether, almost insoluble in cold water.

It is employed as an antipyretic, its action is more rapid than that of acetanilid, yet being less toxic. The dose, although there is no authority upon the subject, would be from 0.1—0.3 Gm. (2-5 grains).

ACET-PARA-AMIDO-SALOL. See Salophen.

ACET-PARA-TOLUID. $C_6H_4 (CH_3) NH. COCH_3$.

Synonym: Para-Tolyl-acetamid.

This is obtained by prolonged interaction between para-toluidin and glacial acetic acid, at boiling temperature. It occurs in colorless crystals of melting-point 149° C. (300.2° F). It is almost insoluble in water, and readily soluble in alcohol.

It is employed as an antipyretic in doses of from 1-2 Gm. (15-30 grains).

ACETYL-PARA-AMIDO-PHENYL SALICYLATE. See Salophen.

ACETYL-AMIDO-ANTIPYRINE. See under Antipyrine.

ACETYL-PARA-ETHOXY-PHENYL-URETHANE. See Thermodin.

ACETYL-PHENYL-HYDRAZINE. See Hydracetine.

ACETYL-TANNIN. See Tannigen.

ACETYL-THYMOL. $C_{12}H_{16}O_2$ or $C_{10}H_{13}O-CH_3CO$.

Synonym: Thymyl Acetate.

This constitutes a colorless liquid of pungent taste, sp. gr. 1.009 at 0° C. Boils at 244.4° C. (472° F.) Employed as an antiseptic.

ACID AGARIC. See Agaricin.

ALPHA-OXY-NAPHTOIC. See Alpha-Oxy-Naphtoic Acid.

ANACARDIC. See Anacardic Acid.

ANGELIC. See Angelic Acid.

ANISIC. See Anisic Acid.

ASEPTIC. See Aseptic Acid.

BETA-PHENYLO-SALICYLIC ACID. See Beta-Phenyl-Salicylic Acid.

BETA-PHENYL-PROPIONIC. See Beta-Phenyl-Propionic Acid.

BOROCITRIC. See Borocitric Acid.

BOROPHENYLIC. See Borophenylic Acid.

BOROSALICYLIC. See Borosalicylic Acid.

CAINCIC. See Caincic Acid.

CAMPHORIC. See Camphoric Acid.

CAMPHORONIC. See Camphoronic Acid.

CATECHU-TANNIC. See Catechu-tannic Acid.

CINNAMYLIC. See Cinnamic Acid.

CRESYLIC. See Cresol.

ACID DI-CHLOR-ACETIC. See Di-Chlor-Acetic Acid.

DI-IODO-SALICYLIC. See Di-Iodo Salicylic Acid.

DITHIOCHLORSALICYLIC. See Dithiochlorsalicylic Acid.

EMBELIC. See Embelic Acid.

FILICIC. See Filicic Acid

GLYCERINO-PHOSPHORIC. See Glycerin-phosphoric Acid.

GYMNEMIC. See Gymnemic Acid.

GYNOCARDIC. See Gynocardic Acid.

HYDRO-CYNNAMIC. See Beta-Phenyl-Propionic Acid.

IODIC, and COMPOUNDS. See Iodic Acid.

ORTHO-AMIDO-SALICYLIC. See Ortho-Amido-Salicylic Acid.

PHENYL-ACETIC. See Phenyl-Acetic Acid.

PHENYLO SALICYLIC. See Phenylo-Salicylic Acid.

SCLEROTIC (Sclerotinic). See Sclerotic Acid.

SOZALIC. See Aseptol.

SOZOIODIC. See Sozoiodic Acid.

SOZOLIC. See Aseptol.

SULPHOTUMENOLIC. See Tumenol.

TETRA-THIO-DICHLOR-SALICYLIC. See Tetra-Thio-Dichlor-Salicylic Acid.

THIOLINIC. See Thiolinic Acid.

TRI-CHLOR-ACETIC. See Tri-Chlor-Acetic Acid.

ADEPS LANÆ.

Synonyms; Lanolin; Adeps Lanæ Hydrosus, U. S. P.; Anasalpin.

"The purified fat of the wool of sheep ......mixed with not more than 30 per cent of water." U. S. P.

The wool of sheep contains a large per cent. of fats (about 45%) which it is necessary to remove before it can be used in manufacturing. These fats consist of a mixture of fatty esters of cholesterin and isocholesterin. The crude wool fat, which is usually obtained by washing the wool with benzine, acetone or some similar solvent and evaporating, is emulsionized with a weak alkaline solution, then separating the creamy mixture in centrifugal machines; the upper layer of fluid contains the cholesterin fats, while the lower layer consists of a soap solution of the impure fatty acids. The upper fluid is drawn off and the cholesterin fats set free by the addition of a solution of calcium chloride; the impure lanolin thus obtained is purified by repeated melting and washing, finally extracting with acetone.

Anhydrous wool-fat is of a pale yellow color, somewhat translucent, melting at 36° C. (96.8° F.), readily soluble in benzine, ether, chloroform, acetone, but only partly soluble in alcohol. When mixed with 30 per cent. of water it constitutes the hydrous wool-fat of the Pharmacopœia.

Hydrous wool-fat occurs as a nearly white ointment-like mass, the surface of which, on standing, becomes of an orange color, due to loss of water. Its melting point is about 40° C. (104° F.); it is miscible with twice its weight of water without losing its ointment-like character. Wool-fat is employed as a base for the preparation of ointments, pomades, creams, etc.

ADHÆSOL.

An antiseptic varnish, recommended as a substitute for Steresol. It contains 350 parts copal resin, 30 parts of benzoin, 30 parts of Tolu balsam, 20 parts of thyme oil, 3 parts of alpha naphthol, and 1,000 parts of ether.

ADONIDIN.

A glucoside of the *Adonis vernalis;* forms a hygroscopic, inodorous, amorphous powder, intensely bitter. It is employed as a cardiac stimulant and mild diuretic in doses of 0.004 to 0.016 Gm. (1-16 to ¼ grain.)

ADONIN.

A glucosidal principle obtained from the herb *Adonis vernalis*. It is a bitter, yellowish-white, hygroscopic powder, soluble in water and alcohol, insoluble in ether. Employed as a cardiac stimulant, being feebly diuretic; dose 0.01—0.06 Gr. (1-6—1 grain.)

ÆSCULIN. $C_{15}H_{16}O_9 + 1\frac{1}{2}$ Aq.

A glucoside obtained from the bark of the horse chestnut. (*Æsculus hippocastanum*.) After precipitating the tannin and coloring matter from the hot aqueous extract by means of alum and ammonia, the filtrate is evaporated to dryness and the residue extracted with alcohol; purified by recrystallization. Æsculin forms inodorous fine needle-like-crystals, of bitter taste, almost insoluble in cold and quite soluble in hot water, its aqueous solution having a strong blue fluorescence. Recommended as an antiperiodic.

ÆTHYL, ÆTHYLEN, ÆTHYLIDEN, ETC., COMPOUNDS.—(See under Ethyl, Ethylene, Ethylidene, Etc.)

AGARIC ACID. See Agaricin.

AGARICIN. $C_{14}H_{27}OH$ (CO_2H) H_2O.

Synonym: Agaric Acid.

The active principle of the fungus *Agaricus albus* or *Polyporus officinalis*, obtained by extraction with alcohol. It forms a white crystalline powder, melting at 140° C. (284° F.). Almost insoluble in cold water, soluble in 130 p. of alcohol.

Agaricin is employed in the treatment of night sweats of consumptives; also in relieving the sudorific effect of the synthetic antipyretics. Dose, 0.01—0.03 Gm. (1-6-½ grain.)

AGATHIN. C_6H_5-N(CH_3)-N : CH-C_6H_4 (OH).

Synonym: Salicyl-x-methyl-phenyl hydrazone.

Agathin is obtained by reaction between salicylic aldehyde and the base methyl-phenyl-hydrazone. This forms colorless crystalline flakes, inodorous and tasteless, insoluble in water, but soluble in alcohol and ether. Its melting point is 76° C. (158° F.)

Agathin is employed as an anti-neuralgic and anti-rheumatic in doses of 0.1—1.5 Gm. (2—8 grains).

AGNINE.

This is probably prepared by distilling wool-fat in a current of superheated steam. It contains a large percentage of free fatty acids (33%)!

AGOPYRIN. See under Antipyrine.

AIROL.

An oxyiodide of bismuth subgallate, patented by the firm of Hoffmann, Traub & Co., of Basel. This compound possesses the absorbent properties of subgallate of bismuth as well as the antiseptic properties of its iodine combination. Airol forms a greenish gray, fine voluminous, inodorous and tasteless powder. Light produces no effect, while moist air causes the powder to turn a red color, with loss of iodine. In contact with water, particularly when heated, the powder undergoes slow decomposition, changing to a red color, with loss of iodine. Dilute alkalies and acids dissolve this compound readily. The formula ascribed is

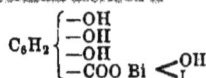

$$C_6H_2\begin{cases}-OH\\-OH\\-OH\\-COO\ Bi\end{cases}<^{OH}_I$$

That is, it is a basic bismuth gallate in which a hydroxyl group is replaced by iodine. Airol is applied as a dusting powder over wounds, sores, etc.

ALANT-CAMPHOR. See under Alantol.

ALANTOL. $C_{10}H_{16}O$.

Elecampane root (*Inula Helenium*) contains, aside from inulin, resinous, waxy and extractive matter; alantol, a liquid stearoptene; alantic anhydride, a crystalline body, and a solid stearoptene called helenin.

Alantol is an aromatic liquid, of a peppermint-like odor, boiling at 200° C. (392° F.). It is employed as an antiseptic in treatment of tuberculous diseases.

Helenin or *Alant-Camphor*. $C_6H_{10}O$.

This occurs in colorless, inodorous crystalline needles, insoluble in water, but soluble in alcohol and ether, melting at 68°—70° C. (155°—158° F.). It is employed as an antiseptic in treatment of malaria, tuberculosis, diarrhœa, etc., in frequent doses of 0.01 Gm. (⅙ grain.)

ALBOLENE.

This is a refined product of petroleum, free from any definite chemical or physiological action; it cannot become rancid, and is introduced as a colorless, odorless base for ointments, cerates, salves, pomades, etc., especially in hot climates.

Albolene liquid is a colorless, odorless, tasteless fluid, with a specific gravity of **0.865** at 60° F., obtained from petroleum and afterwards specially treated. It does **not** saponify nor become rancid, neither is it decomposed by acids or alkalies. It can be vaporized in a brine bath at zero, indicating the absence of solid paraffin, and is free from petroleum ethers. Its lightness makes it very diffusible **as** vapor or as a **solvent** for drugs used in oil for spraying the nasopharyngeal **passages.**

ALIZARIN-YELLOW-C.　See Gallacetophenon.

ALLYL-MUSTARD OIL.　See Oleum Sinapis Volatile.

ALLYL-SULPHO-CARBAMIDE.　See Thiosinamine.

ALLYL-SULPHO-UREA.　See Thiosinamine.

ALLYL-THIO-UREA.　See Thiosinamine.

ALLYL TRIBROMIDE.　$(C_3H_5Br_3)$

A colorless or slightly yellowish liquid, insoluble in water, soluble in ether. Recommended as sedative and anodyne in hysteria, asthma, whooping cough, etc.; in doses of 2 to 4 drops dissolved in ether, hypodermically.

ALPHA-CREOSOTE.

This artificial product is prepared by mixing together the several constituents as found in normal creosote in such proportions that it contains 25 per cent. crystalline guaiacol.

ALPHA-GUAIACOL.

A crystalline, synthetic guaiacol. This name is applied to distinguish it from the commercial guaiacol.

ALPHOL.　See Betol.

ALUMINUM-ACETO-TARTRATE.

This occurs in yellowish granules, with acid, astringent taste, slightly soluble in water. It is an astringent and disinfectant in nasal and laryngeal affections. Apply in ½ to 2% solutions, or as a snuff with 2 parts of boric acid.

ALUMINUM-AMMONIUM-SALICYLATE.　See Salumin (soluble).

ALUMINUM-BASIC TANNATE.　See Tannal (insoluble).

ALUMINUM-BETA-NAPHTHOL-DISULPHONATE.　See Alumnol.

ALUMINUM-BOROFORMATE.

It is thus made: Mix 2 parts formic acid, 1 part boric acid and 6 or 7 parts of water, and in the mixture dissolve fresh precipitated alumina; filter, allow to crystallize; or it may be employed in a solution of sp. gr. 1.064 (10%) or sp. gr. 1.110 (20%), reduced by careful evaporation. It is employed as a mild antiseptic and astringent.

ALUMINUM-BORO-TANNICO-TARTRATE.　See Cutal.

ALUMINUM-BORO-TARTRATE.　See Boral.

ALUMINUM-GALLATE (Basic).　See Gallal.

ALUMINUM-PARA-PHENOL-SULPHONATE.　See Sozal.

ALUMINUM-POTASSIUM-PARA-PHENOL-SULPHONATE.

$Al_2K_2(C_6H_4(OH)SO_3)_6.$

This compound is obtained by saturating para-phenol sulphonic acid with a solution of potassium-aluminate. It forms colorless crystals, soluble in water. Its properties are those of an antiseptic and astringent; it is employed chiefly as a wash for indolent ulcers, etc.

ALUMINUM-POTASSIUM SALICYLATE.

This double salt is produced by a process recently patented by Atheustedt, in which hot potassium acetate is made to act on aluminum salicylate. It contains a high proportion of alumina, and is said to be a powerful astringent as well as possessing high antiseptic properties.

ALUMINUM SALICYLATE.　See Salumin (insoluble.)

ALUMINUM TANNATE (Basic.)　See Tannal.

ALUMINUM TANNIC-TARTRATE.　See Tannal (soluble.)

Alumnol is employed as an antiseptic and astringent. As a wash for purulent surfaces, it is employed in from 1 to 5% solution ; in 10 to 20% solutions it serves as a cautery. Mixed with powdered talc or starch (10 to 20%), it forms an astringent dusting powder.

ALSTONIN.

An alkaloid obtained from the bark of the *Alstonia constricta* (Apocynaceæ.) Alstonin forms white, crystalline needles, insoluble in water; soluble in alcohol, ether and chloroform. To hot water it imparts an intensely bitter taste.

ALUMNOL. $(C_{10}H_6 (OH) SO_3)_3Al$.

Synonym : Beta-Naphthol-Disulphonate of Aluminum.

Alumnol is obtained by reaction between the barium compound of beta-naphthol-disulphonic acid and aluminum sulphate. It forms a colorless powder, readily soluble in water and glycerine, but only slightly soluble in alcohol. Aqueous solutions are incompatible with alkaline solutions, the hydrate of alumina being precipitated. Likewise it precipitates albuminoid and gelatinous bodies from solution, the precipitate being soluble in excess of albumen or gelatin. Alumnol should not be brought in contact with ammoniacal compounds.

AMIDO-ACET-PHENETIDIN. See Phenocoll.

AMIDO-ANTIPYRINE. See under Antipyrine.

AMIDO-GUAIACOL.

Aceto-o-anisidin, on treatment with nitric acid, yields nitro-acet-o-anisidin, which, on being boiled with alkalies, undergoes saponification, yielding an alkali salt of nitro-guaiacol. By the action of zinc and hydrochloric acid or other reducing agents, this nitro-guaiacol is converted into amido-guaiacol. This base melts at 184° C. (363.2° F.); its hydrochloride at 242° C. (467.6° F.) The salts of amido-guaiacol are employed in the preparation of colors and as medicinal agents.

AMIDO-SUCCINAMIDE. See Asparagin.

AMINOL.

A liquid disinfectant, possessing a strong, fishy odor, of alkaline reaction. One liter of aminol is said to contain 1.52 Gm. of calcium hydrate, 3.516 Gm. of sodium chloride, and 0.29 Gm. of trimethylamine.

AMMONIUM EMBELICUM. $C_9H_{13}O_2 . NH_4$.

This is the ammonium salt of embelic acid, the latter being prepared from the *Embelia Ribes,* Burm. (*Myrsineæ.*)
It is a brick-red powder, readily soluble in diluted alcohol.
It is employed as a tænifuge, in doses from 0.18 Gm. (2.8 grains) for children; to 0.4 Gm. (6. grains) for adults.

AMMONIUM-ICHTHYOL-SULPHONATE. See Ichthyol.

AMMONIUM PERSULPHATE. $(NH_4)_2S_2O_8$.

Recommended as an antiseptic for preserving food. It occurs in small colorless crystals, soluble in water. The solution evolves active oxygen when heated. It can be used in place of permanganate.

AMMONIUM SALICYLATE. $C_6H_4 (OH) COONH_4$.

This compound is obtained by neutralizing salicylic acid with ammonium carbonate, evaporating and crystallizing. It forms a white crystalline powder, of sweet taste and very soluble in water.
Ammonium Salicylate is recommended as an expectorant, the dose being the same as the other salicylates.

AMMONIUM-SULPHO-ICHTHYOLICUM. See Ichthyol.

AMMONOL.

A proprietary remedy recommended as an antipyretic and analgesic. It is also claimed to possess antiseptic properties. Dose 0.3 to 1.3 Gm. (5 to 20 grains.)

AMYGDOPHENIN.

A paramidophenol derivative in which one of the hydrogen atoms of the amide group, is replaced by an ester of amygdalic acid, and the hydrogen atom in the hydroxyl group is replaced by ethyl carbonate. It has been commended as an anti rheumatic by Dr. Stueve, who administered it in doses of 1 Gm. (15 grains) from one to six times daily, in a powder or tablet. Amygdophenin occurs as a grayish-white, light, crystalline powder, very difficultly soluble in water.

AMYL-ALCOHOL, TERTIARY. See Amylene Hydrate.
AMYLENE. See Pental.
AMYLENE HYDRATE. $(CH_3)_2.C_2H_5.C.OH.$

Synonyms: Tertiary Amyl Alcohol; Amylenum Hydratum; Dimethyl-Ethyl-Carbinol.

This is one of the eight possible alcohols of the general formula, $C_5H_{11}OH$; it is prepared by the action of sulphuric acid upon amylene (C_5H_{10}), the latter being obtained by the action of dehydrating agents on isobutylcarbinol.

Amylene Hydrate is a colorless, limpid fluid of peculiar penetrating odor, similar to that of peppermint. Its specific gravity is 0.815, and boiling point is between 99°-103° C. (210°-217° F.) It dissolves in eight parts of water, and is miscible with alcohol, ether, glycerine and the fatty oils.

Tertiary Amyl Alcohol is employed as a hypnotic in doses of 2 to 4 Gm. (30 to 60 grains.)

AMYLOCARBOL.

A mixture of nine parts of phenol; green soap, 150 parts; amylic alcohol, 160 parts, and water sufficient to make 1,000 parts.

AMYL-VALERIANATE. $C_5H_9O_2\ C_5H_{11}$.

Synonyms: Apple Oil; Iso-Amyl-Valerianate.

This well known ester is a colorless ethereal liquid, which boils at 194° C. It possesses a solvent action on cholesterin; also exerts a specific stimulating and sedative action on the liver in gallstone colic. Usually administered in gelatin capsules containing 0.15 Gm. (2.3 grains.)

ANACARDIC ACID. $C_{22}H_{32}O_3$.

A crystalline principle obtained from the *Anacardium occidentale* (cashew nut). It forms hygroscopic crystalline masses, readily soluble in alcohol, melting at 26° C. (78.8° F.) Anacardic acid is used as an anthelmintic, usually in the form of the ammonium salt.

ANADOL.

A proprietary antipyretic.

ANÆSTHYL.

A mixture of 5 parts of ethyl chloride and 1 part of methyl-chloride. Employed as a local anæsthetic spray

ANAGYRINE HYDROBROMATE $C_{14}H_{18}N_2O_3HBr_2$.

This is a salt of the alkaloid obtained from the seed of *Anagyris fœtida.* It occurs as small, white shining scales, which are soluble in water and alcohol, and melt between 20° and 266° C. Physiological investigations by Hardy and others have proved anagyrine to be toxic; therapeutic data are wanting.

ANALGENE. $C_9H_5\text{-}OC_2H_5\text{-}NH.\ COC_6H_5\text{-}N.$

Synonyms: Benzanalgene; Ortho Ethoxy Ana-Mono-Benzoyl-Amido-Chinolin; Ethoxy-Ana-Benzoyl-Amido-Chinoline; Quinalgene.

This body is obtained by introducing an ethyl and amido group into ortho-oxychinolin; into the resulting ortho-oxyethyl-amido-chinolin the benzoyl group is introduced by the action of benzoyl chlorid.

Analgene forms white, tasteless crystals, melting at 208° C. (406.4° F.), insoluble in water, readily soluble in alcohol.

This compound is employed as an anti-neuralgic in doses of 0.5 to 1.0 Gm. (8 to 15 grains.)

ANALGESIN. See Antipyrin.

ANALGIA.

Proprietary analgesic and antipyretic.

ANASALPIN.

Synonym for anhydrous wool fat.

ANASPALINE.

This consists of a mixture of wool-fat, with about 25% of petrolatum.

ANDUNEA.

A proprietary analgesic.

ANEMONIN. $C_{10}H_8O_4$.

This is the active principle of the herbs *Anemone pulsatilla, A. pratensis* and *Ranunculus acer.*

It occurs in colorless, crystalline needles, which melt at 152° C. (305.6° F.), and readily dissolve in warm alcohol and the oils while being almost insoluble in water. Anemonin is employed in treatment of whooping cough, bronchitis and asthma, in doses of 0.004 to 0.03 Gm. (1-16 to ½ grain.)

ANETHOL. $C_3H_5-C_6H_4-O-CH_3$.

Synonyms: Para-Allyl-Phenyl-Methyl-Ether; Anise Camphor.

Anethol constitutes the main constituent of oil of anise. It occurs in colorless crystals, melting at 22° C. (71.6° F.) and boiling at 234° C. (453° F.), soluble in alcohol and the oils, but insoluble in water.

Anethol is employed as an antiseptic, also as a flavoring constituent in liquors, etc.

ANGELIC ACID. $CH_3(CH)_2 CH_2COOH$.

A principle prepared from the root of *Angelica Archangelica.* It forms monoclinic prisms, of spicy odor, soluble in alcohol, ether and hot water ; melts at 45° C. (113° F.) Used as an aromatic tonic.

ANGIONEUROSIN. See Nitroglyc·rin.

ANHALONINE. $(C_{12}H_{15}NO_3.)$

An alkaloid from the Mexican cactacea *Anhalonium lewinii*, which contains it in both amorphous and crystalline forms. The crystalline base melts at 85° C, (185° F.), soluble in alcohol and ether ; the hydrochlorate is deliquescent and very soluble in water. Anhalonine is of value in treatment of angina pectoris, asthmatic dyspnœa and pneumothorax.

ANHYDRO-GLUCO-CHLORAL. See Chloralose.

ANISE CAMPHOR. See Anethol.

ANISIC ACID. $C_6H_4(OCH_3)COOH$.

Synonym: Paramethoxy-Benzoic Acid.

This is an isomer of methyl salicylic acid, obtained by oxidation of anethol, a constituent of oils of anise and fennel.

It forms colorless prismatic crystals, which melt at 180° C. (356° F.) It is insoluble in cold water, but very soluble in alcohol.

Employed externally it possesses antiseptic properties. Internally it exerts antipyretic and anti-rheumatic properties. It is usually administered as sodium salt.

SODIUM ANISATE. This is obtained by neutralizing anisic acid with sodium carbonate or bicarbonate. The commercial salt constitutes a hygroscopic, crystalline powder of less disagreeable taste than the acid.

The dose is 1 Gm. (15 grains.)

PHENYLESTER OF ANISIC ACID. $C_6H_4(OCH_3)CO_2C_6H_5$. This compound bears the same relationship to anisic acid as salol does to salicylic acid. It is obtained by the action of phosphorus pentachlorid on a mixture of anisic acid and phenol.

It occurs as colorless crystals, which melt at 75° C. (167° F.), insoluble in water, but very soluble in alcohol and ether.

It is employed in the treatment of neuralgia and rheumatism, in doses from 0.5 to 1.0 Gm. (8 to 15 grains.)

ANNIDALIN.

Synonym: Di-thymol-triiodide.

This occurs as a reddish-brown powder, which is readily decomposed by heat and light; it is insoluble in water and slightly soluble in alcohol. Annidalin is usually applied locally as a substitute for iodoform and aristol. It is used as pure powder or as a 10 per cent. dilution.

ANODYNIN (Anodin.) See Antipyrin.

ANTALGIA.

A proprietary antipyretic and analgesic.

ANTHRAROBIN. $C_6H_4 < \begin{array}{c} C\,(OH) \\ | \\ CH \end{array} > C_6H_2\,(OH)_2$.

Synonyms: Dioxyanthrol; Desoxy-Alizarin; Leuko-Alizarin.

A phenol derivative related to chrysophanic acid, obtained by the reduction of alizarin.

Anthrarobin is a yellowish white powder, insoluble in water, but very soluble in aqueous solutions of the alkalies and alkaline earths. These alkaline solutions rapidly turn green, then blue, through absorption of oxygen from the air, alizarin being reformed.

It is employed as a **substitute for chrysarobin in skin diseases**, usually as a 10 to 20% ointment.

ANTI-BACILLARE.

A remedy for phthisis, consisting of a mixture of **creosote, tolu balsam, glycerin, codeine and sodium arsenate.** Dose not known.

ANTIBACILLIN.

A proprietary disinfectant.

ANTIDIPHTHERIN.

A sterilized solution containing cultures of diphtheria bacillus, in addition 0.2% of ortho-cresol and some glycerin. It occurs in commerce in two concentrations, one double and the other four times the strength of the original culture fluid. The stronger solution is employed for painting the affected parts, while the weaker is used for hypodermic injection (½ cc.)

ANTIDIPHTHERIN (KLEBS).

This preparation is designated by the letters A. D., and a given volume of it is 10 times as strong as the original culture fluid. It is obtained from this, after removal of the bacilli, by precipitation with alcohol.

ANTIDIPHTHERITIKON.

Bokai's mixture of Oil of Birch (5), **Oil of Beech (3).** Alcohol (90), Potassium Carbonate (1), and Potassium Sulphide (5). **Used as a diphtheria remedy.**

ANTIDOLOR.

A proprietary anodyne.

ANTIFEBRIN. See Acetanilid.

ANTIFETOR.

A proprietary deodorizing powder.

ANTIHYDROPIN.

A crystalline body supposed to be derived from the common cockroach, recommended as a diuretic in daily doses of 0.6 to 1.3 Gm. (10-20 grains.)

ANTIKAMNIA.

A secret remedy employed as an antipyretic and analgesic. Various analyses have shown the presence of acetanilide, sodium bicarbonate and caffeine. Its dose is given as 0.3 to 0.6 Gm. (5 to 10 grains.)

ANTIKOL.

A powder said to be a mixture of acetanilide (75p.), sodium bicarbonate (17.5p.) **and** tartaric acid (7.5 p.) (Goldmann). Dose, 3 to 10 grains.

ANTINERVIN.

A powder consisting of acetanilide (50 p.), ammonium bromide (25 p.), **and salicylic acid** (25 p.) (Squibb).
It is employed as an antinervine and antipyretic, in doses of 0.5 Gm. (8 grains).

ANTINONNIN. See under Cresol.

ANTINOSIN.

The sodium salt of nosophen.

ANTIPYONIN.

Synonym: Sodium Tetraborate.
A polyborate of sodium, recommended as a remedy in inflammation of the cornea and conjunctiva. It is a fine white powder of greasy feeling, freely soluble in water, and devoid of caustic action.

ANTIPYRALGOS.

A proprietary antipyretic and anodyne.

ANTIPYRINE. $C_{11}H_{12}N_2O$.

Synonyms: Analgesine; Anodynin; Parodyn; Oxy-Dimethyl-Chinizin; Phenazon; Phenyl-Dimethyl Pyrazolon; Phenylon; Salazolon; Sedatin; Di-Methyl-Phenyl-Pyrazolon; Methozine; Pyrazine; Pyrazolin.

This is a synthetical base, obtained by the action of aceto-acetic-ester on phenylhydrazine, the resulting phenyl-methyl-pyrazolon being methylated.

Antipyrine occurs in colorless and odorless crystals, which melt between 112°-113° C. (233 6°-235 4° F.), and are readily soluble in water and alcohol.

Its solutions are turned a green color on addition of nitrous acid (Sp. Etheris Nitrosi), and a deep red color on addition of ferric chloride (Tr. Ferri Chloridi). Because of its strongly basic properties, antipyrine presents a number of incompatibles.

The following is a list of the more important of these :

Acidum Hydrocyanicum Dil.
Acidum Tannicum. All galenical preparations containing tannin form insoluble precipitates.
Acidum Carbolicum ; either one precipitates the other from solution.
Chloral Hydrate and Butyl Chloral.
Ferri Sulphas.
Ferric Salts in solution (red color). Frequently dispensed.
Liquor Arsenii et Hydrargyri Iodidi ; insoluble precipitate.
Mercurous and Mercuric Chloride.
Nitrites in solution (green color). Frequently dispensed.
Sodii Bicarbonas.
Tinctura Iodi ; insoluble precipitate.
When triturated with Chloral, Phenyl-Urethane, Beta-Naphthol or Sodium Salicylate, it forms a pasty or liquid mass.
Antipyrine increases the solubility of Caffeine and Quinine Salts in water
Internally it is employed as an antipyretic, antirheumatic and antineuralgic in doses of 1 to 2 Gm. (15 to 30 grains) for adults, and 0.2 to 0.5 Gm. (3 to 8 grains) for children. Externally, antipyrine is used as an antiseptic and hæmostatic.

DERIVATIVES AND ALLIED COMPOUNDS.

Antipyrine, being a basic body, readily unites with acids to form salts, a number of which have been introduced into medicine.

SALIPYRINE, or *ANTIPYRINE SALICYLATE (SALAZOLON, SALIPYRAZOLON)* $(C_{11}H_{12}N_2O.{}_7H_6O_3)$, is obtained by interaction between antipyrine and salicylic acid; 57.7 parts of the former and 42.3 parts of the latter being heated together on a water bath, the resulting oily-like fluid which solidifies on cooling is crystallized from alcohol. It occurs as a crystalline, inodorous powder, melting at 92° C. (197.6° F.), soluble in 200 parts of cold and 25 parts of boiling water, very soluble in alcohol and ether. Salipyrine exerts the combined effects of antipyrine and salicylic acid, being employed in the treatment of acute and chronic rheumatism, rheumatic sciatica and influenza, in doses of 1 to 2 Gm. (15 to 30 grains.)

AGOPYRIN. A secret remedy which appears in the form of tablets containing ammonium chloride 1-3 grain, cinchonine sulphate 1-3 grain and salicin 4 grains.

ACETYL AMIDOANTIPYRINE is the acetyl derivative of amido-antipyrine, which in turn is obtained by the reduction of iso-nitroso-antipyrine. This latter body results from the action of nitrous acid on antipyrine. It forms a crystalline powder which melts at 197° C. (386.6° F.), its properties being similar to those of antipyrine.

AMIDOANTIPYRINE occurs in yellow needles which melt at 109° C. (288° F.) It is obtained by the reduction of isonitroso-antipyrine, which results from the action of nitrous acid on antipyrine.

BUTYL-HYPNAL is a combination of butyl chloral and antipyrine, similar to that of the latter with chloral hydrate (see Hypnal.) It forms colorless needles, melting at 70° C. (158° F.), soluble in 30 parts of water and readily soluble in alcohol.

DICHLORALANTIPYRINE $(C_{11}H_{12}N_2O+2CCl_3CH(OH)_2)$ is obtained by triturating 94 parts of antipyrine with 165.5 parts of chloral hydrate until a mass is obtained, which is crystallized from hot water. Its medicinal properties are like those of monochloralantipyrine.

IODOPYRINE, or *IODOANTIPYRINE* $(C_{11}H_{11}IN_2O)$, forms colorless needles, which melt at 160°C. (320° F.), being only slightly soluble in water. Iodopyrine has the action of an iodide in addition to that of antipyrine; its dose being 0.5 to 1.5 Gm. (8 to 23 grains).

MIGRANIN This is an antipyrine preparation, the composition of which is, according to various analyses, antipyrine 89.4%, caffeine 8.2%, citric acid 0.56% and moisture 1.84%. It is recommended as a specific in the treatment of migraine, and is also employed in relieving the headache of influenza. The dose is given as 1 Gm. (15.5 grains).

PARA-METHOXY-PHENYL-DIMETHYL-PYRAZOLON, or *PARA-METHOXY-ANTIPYRINE*. Obtained by methylating the product of the reaction between p-methoxy-phenyl-hydra-zine and acetoacetic ester. It forms crystals which melt at 82° C. (170.6° F.), being readily soluble in water and alcohol. The *ethoxy* compound melts at 91° C. (195.8° F.)
Both of these compounds possess antipyretic and antineuralgic properties, being weaker, however, than antipyrine.

MONOCHLORALANTIPYRINE, or *HYPNAL* $(C_{11}H_{12}N_2O. CCl_3CH (OH)_2)$, is obtained by triturating together 188 parts of antipyrine and 165.5 parts of chloral hydrate until liquefaction takes place ; the oily-like fluid is then dissolved in hot water and set aside to crystallize. Hypnal forms colorless crystals, which melt at 67.5° C. (154° F.), and are readily soluble in warm water. It is employed as a hypnotic and analgesic in doses of 1 to 2 Gm. (15 to 30 grains).

NAPHTHOPYRINE (Beta-Naphthol-Antipyrine) is a molecular compound of beta-naphthol and antipyrine, obtained by trituration.

PHENOPYRINE is prepared from equal parts of phenol and antipyrine. It is an oily, colorless fluid, free from odor, insoluble in cold water.

PYROGALLOPYRINE is obtained by reaction between pyrogallol and antipyrine.

RESOPYRINE ($C_{11}H_{12}N_2O+C_6H_4 (OH)_2$) is obtained by reaction between antipyrine and resorcin in solution in molecular proportions. This is obtained by crystallization in fine rhombic crystals, which are insoluble in water and soluble in 5 parts of alcohol. Nothing definite is known concerning its therapeutic properties.

ANTIPYRINE MANDELATE. See Tussol.

ANTIPYRINE PHENYL-GLYCOLATE. See Tussol.

ANTIPYRINE SALICYLATE. See Salipyrin, under Antipyrin.

ANTIPYRINE-SALOL.

Obtained by fusing together equal quantities of salol and antipyrine, heating until the fluid turns a brown color and remains fluid on cooling. It is recommended as an antiseptic, also as a valuable hæmostatic in uterine hemorrhages, in which case it is applied by means of cotton tampons.

ANTIRHEUMATICUM.

A mixture of methylene blue and sodium salicylate.

ANTIRHEUMATIN.

This compound of sodium salicylate and methylene blue (q. v.) forms blue, prismatic crystals, soluble in water and alcohol. Recommended as an antirheumatic, in doses of 0.06 to 0.09 Gm, (1 to 1½ grains).

ANTISEPSINE. C_8H_4Br NH-CH_3CO.

Synonyms: Asepsine; Para-Brom-Acetanilid ; p-Mono-Brom-Phenyl-Acetamid

Asepsine is obtained by adding bromine, in molecular proportions, to a solution of acetanilid in glacial acetic acid ; the white precipitate formed is recrystallized from alcohol. It forms colorless crystals, which melt between 165° and 166° C. (329° to 330.8° F.) It is but slightly soluble in water, more so in alcohol. Its properties are those of an antipyretic, in doses of 3 to 10 grains; it is also of value in treatment of muscular rheumatism and neuralgia. Not to be confounded with a mixture of zinc sulphate, zinc iodide, thymol and boric acid called " Antiseptin."

ANTISEPTIN.

A mixture containing about 80 parts of zinc sulphate, 2 parts of thymol and 18 parts of boric acid. Recommended as an antiseptic. This should not be confounded with antisepsin or antiseptol.

ANTISEPTOL. (Chemical Formula Uncertain).

Synonyms: Antiseptolum ; Cinchonin-Herapathit ; Cinchonin Iodosulphate.

To a solution of 25 parts of cinchonin sulphate, in 2,000 parts of water, is added a solution of 10 parts of iodine and 10 parts of potassium iodide in water, the precipitate is collected, washed and dried. This constitutes a red-brown powder, which is insoluble in water, but very soluble in alcohol and chloroform. It contains about 50% of iodine.

Antiseptol is employed as a substitute for iodoform.

ANTISPASMIN. $C_{23}H_{26}NO_8Na+3C_6H_4 (OH) COONa$.

This is a double salt of narceine sodium and salicylate of sodium. It forms a white, slightly hygroscopic powder, which dissolves readily in water. The compound contains about 50% of narceine and has an alkaline reaction. Antispasmin is a hypnotic and sedative, being administered in doses of 0.01 to 0.1 Gm. (⅙ to 1½ grains).

ANTISTREPTOCOCCIN.

A serum preparation used as a remedy against erysipelas.

ANTITETRAIZIN.

A derivative of quinine, recommended by Zambeletti in treatment of neuralgia, influenza, etc., in doses of 0.2 to 0.25 Gm. (3 to 4 grains).

ANTITHERMAL.

A proprietary febrifuge.

ANTITHERMIN. $CH_3C(C_6H_5N_2H)C_2H_4COOH$.

Synonym: Phenylhydrazine-Lævulinic Acid.

This compound is obtained by interaction between a solution of phenylhydrazine in acetic acid and lævulinic acid. It occurs in colorless, tasteless crystals, which melt at 108° C. (226.4° F.), almost insoluble in cold water.

Antithermin is employed as an antipyretic in pulmonary phthisis **and morbus** Brightii, the dose being 0.2 Gm. (3 grains).

ANTITOXINE. See Diphtheria Antitoxine.

ANTITOXINE.

A proprietary antipyretic, not to be confounded with the generic term "Antitoxine." the blood serum of immunized animals.

APIOL. $C_{12}H_{14}O_4$.

A stearoptene obtained from the fruits of *Petroselinum Sativum.* The alcoholic extract of the fruit is reduced to extractive consistence and the extract washed with ether, whereby the apiol goes into solution and on evaporation of the ether crystallizes. The so-called *liquid apiol* is merely an alcoholic extract of the parsley fruit.

Apiol forms colorless needles, of feeble parsley-like odor, melting at 32° C. (89.6° F.), insoluble in water, very soluble in alcohol, ether, fixed and volatile oils.

It is employed as an antiperiodic and against dysmenorrhœa. Dose 0.25 Gm. (4 grains).

APOCODEIA HYDROCHLORATE. $C_{18}H_{19}NO_2$-HCl.

Synonym: Apocodeinum Hydrochloricum.

This is prepared from codeine in a manner analogous to the manufacture of apomorphine from morphine. Apocodeine hydrochlorate forms an amorphous, yellowish powder, soluble in alcohol and water. Its properties are similar to those of apomorphine, it being employed as an expectorant in doses of 0.06 to 0.08 Gm. (1 to 1¼ grains).

APOLYSINE.

Synonym: Monophenethydin.

A substance approaching phenacetin very closely in its chemical composition, etc. It appears as a yellowish-white crystalline powder, soluble in 50 parts of cold water and 25 parts of hot, glycerin, concentrated nitric and sulphuric acids, etc. In its physiological action, etc., it resembles phenacetin very closely, lowering the tempera ture, allaying pain, and is said to be free from the unpleasant after-effects of that substance (phenacetin). It appears to be absolutely non-toxic, no injurious effects being noted when given to rabbits and other warm-blooded animals by hypodermic injections of 3.50 Gm. **of a** 1% solution to each kgm. of weight.

APYONIN.

A **yellow, crystalline** powder, introduced as a rival to auramine **for use in ophthalmic surgery.** It is slightly **soluble** in water and readily in alcohol.

ARBUTIN. $(C_{12}H_{16}O_7)_2+H_2O$.

A glucoside obtained from the leaves of the bearberry, *Arctostaphylos uva ursi.*

It occurs in colorless, silky needles, which melt at 170° C. (338° F.), soluble in 8 parts of cold water and 16 parts of alcohol. Arbutin is employed in diseases of the kidneys and urinary tract, being given in doses up to 5 Gm. (75 grains).

ARECOLIN. $C_8H_{13}NO_2$.

A liquid alkaloid obtained from the betel-nut *(Areca catechu).* It is a strongly alkaline liquid, miscible with water, alcohol or ether; boiling at 220° C. (428° F.)

Arecolin is employed as an anthelmintic, in doses of 0.003 to 0.004 Gm. (1-20 to 1-16 grain). Great care should be observed in its administration, as it is a powerful heart poison.

The *hydrochlorate* of *arecolin,* a **soluble crystalline salt, melting** at 167° C., **is also** employed for the **same** purposes as **the above.**

ARGENTAMIN.

An antiseptic, employed in gonorrhœa. It is a solution of silver phosphate **in** aqueous solution of æthylendiamine. In the preparation of this the manufacturers have sought to present an antiseptic, which does not precipitate albumen, held in solution in a non-corrosive and non-toxic solvent. It has been found that the antiseptic **power** of strong alkaline solvents is greater than simple aqueous solutions, since the alkalies dissolve the membrane of the micro-organism; as organic bases adopted to **this purpose** are æthylendiamine and alkyl derivatives, piperazin, etc. As antiseptics, which, in conjunction with these organic bases, do not precipitate albumen, are phenol, cresol, thymol, naphthol, guaiacol and silver salts. The solutions are prepared thus, after the patent : 10 parts of æthylendiamine are dissolved in 500 parts of water, adding 10 parts of freshly dissolved cresol. Where creosote or guaiacol is employed, it is better to use a larger amount of the base (æthylendiamine). For the preparation of the silver solutions 10 parts of silver phosphate (nitrate or chloride) are added slowly, with constant stirring, to a solution of 10 parts of the base in 100 parts of water.'

ARGONIN.

A compound of silver, casein and alkali, prepared by adding a solution of the sodium compound of casein to a solution of silver nitrate, and precipitating the newly formed body by the addition of alcohol. The resulting white powder must be free from nitric acid and alkali. Argonin is insoluble in cold, but readily soluble in hot water ; its solutions must be kept away from the light. It is incompatible with acids. Employed as a powerful bactericide.

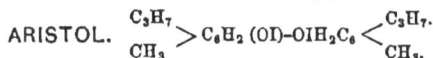

ARISTOL. $\dfrac{C_3H_7}{CH_3} > C_6H_2(OI)-OIH_2C_6 < \dfrac{C_3H_7.}{CH_3.}$

Synonyms: Annidalin ; Di-Iodo-Dithymol : **Di-Thymol-Iodide.**

To a solution of 5 parts of thymol and 1.2 parts of sodium-hydrate in 10 parts of water, add gradually under constant stirring, a solution of 6 parts of iodine and 9 parts of potassium iodide in 10 parts of water. The precipitate is washed with water and dried at low temperature. Aristol forms a pale chocolate colored amorphous powder, insoluble in water and glycerin, slightly soluble in alcohol, readily in ether and collodion; light and heat cause its decomposition. Aristol contains 45.8 per cent. of iodine. This compound was introduced as a substitute for iodoform, possessing the advantage of being odorless. Ointments containing it are usually prepared of the strength of 5 to 10 per cent.; other forms of applying it are solutions in oils, ether and collodion.

AROPHENE. A proprietary dental anæsthetic.

ASAPROL. $(C_{10}H_6. b\ OH. a\ SO_3)_2\ Ca+3H_2O.$

Synonyms : Beta-naphtol-a-mono-sulphonate of calcium; Abrastol.

This is prepared by saturating an aqueous solution of Beta-naphtol-a-monosulphonic acid with calcium carbonate, evaporating and crystallizing the salt. Asaprol forms a colorless to pale reddish inodorous powder, which is soluble in 1½ parts of water and 3 parts of alcohol,

It is employed as an antipyretic and analgesic in doses of 3 to 12 grains four or five times daily. Incompatibles are sulphates, bicarbonates, iodides, antipyrine and quinine.

ASBOLINE.

Prepared from pine root. Yellowish oil consisting principally of pyrocatechol and its homologues. Antitubercular.

ASEPSINE. See Antisepsine.

ASEPTIC ACID.

This should not be confounded with aseptol, asepsin or antisepsin. Aseptic acid is an aqueous solution of 5 Gm. boric acid in 1000 Gm. of hydrogen peroxide (5%) with or without the addition of 3 Gm. of salicylic acid. (Thoms.)
The properties of this solution are antiseptic.

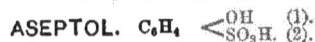

ASEPTOL. $C_6H_4 < \dfrac{OH\ (1).}{SO_3H.\ (2).}$

Synonyms: Acidum Sozolicum ; Sozolic Acid ; Ortho-Phenol-Sulphonic Acid ; Ortho-Sulpho-Carbolic Acid.

Equal parts of concentrated sulphuric acid and phenol are mixed, the two liquids kept at as low a temperature as possible, otherwise, by higher temperature the para-acid forms instead of the ortho. After standing several days, it is poured into water and neutralized with barium carbonate ; the barium sulphate is filtered off, the filtrate containing the barium salt of sozolic acid. The barium is removed from this compound by careful addition of sulphuric acid.

Aseptol is a 33 1-3 per cent. solution of ortho-phenol-sulphonic acid, its specific gravity being 1.155; it possesses a feeble odor resembling phenol. On long standing it gradually goes over into the para compound. Aseptol is employed as an antiseptic wash in 10 per cent. solution. Its solutions in glycerin, oil or alcohol are inactive. It is administered internally as an anti fermentive in like doses as salicylic acid.

ASPARAGIN. $C_9H_3(NH_2)<{}^{CONH_2}_{COOH}+H_2O$

Synonyms: Asparamid ; Amido-Succino-Amide.

A crystalline principle which occurs in asparagus and marshmallow root, being obtained from the latter by evaporating the aqueous extract to a concentrated volume and crystallizing.

Asparagin forms colorless crystals, which are only slightly soluble in cold water and alcohol.

It is employed as a diuretic in doses of 0.05 to 0.1 Gm. (0.7 to 1.5 grains).

ASPARAMID. See Asparagin.

ASPIDOSPERMIN. $C_{20}H_{30}N_2O_2$.

An alkaloid isolated from the bark of *Aspidosperma Quebracho*. It occurs in colorless crystals, which are insoluble in water and soluble in about 48 parts of alcohol. It is employed in treatment of asthma, dyspnœa, emphysema, etc., in doses of 0.016 to 0.03 Gm. (¼ to ½ grain).

ATHANON.

A proprietary disinfectant.

AURAMINE. See Pyoktaninum Aureum.

BACILLIN.

A proprietary deodorizer and disinfectant.

BENZACETIN.

Synonym: Acet-Amido-Methyl-Salicylic Acid.

This compound forms colorless crystals, melting at 205° C. (401° F.), and almost insoluble in water. Recommended as an antineuralgic, in doses of 0.5 to 1. Gm. (8 to 15 grains).

BENZANALGENE. See Analgene.

BENZANILIDE. $C_6H_5NH CO C_6H_5$.

Synonym: Benzoyl-Anilide.

Obtained by the action of benzoyl chloride on aniline in the presence of caustic soda. It forms colorless crystals, insoluble in water, soluble in alcohol, melting at 163° C. (325.4° F.)

Benzanilide is employed as an antipyretic, suitable for children, the dose being 0.1 to 0.5 Gm. (1.5 to 8 grains), according to age.

BENZONAPHTHOL. $C_6H_5CO_2-C_{10}H_7$.

Synonyms: Beta-Naphthol Benzoate ; Benzoyl-Beta-Naphthol.

This compound is analogous to betol (naphtho-salol), being a naphthyl ester of benzoic acid $(C_6H_5CO_2H)$, and is obtained by the action of benzoyl chloride on beta-naphthol. Benzonaphthol occurs in crystalline needles, or powder, inodorous, tasteless, insoluble in water, soluble in alcohol and chloroform, melts at 110° C. (230° F.) Employed as an intestinal antiseptic in doses of 0.25 to 0.5 Gm. (4 to 8 grains), being split up in the intestines into beta-naphthol and benzoic acid.

BENZO PARA-CRESOL. $C_6H_4(CH_3)(CO-C_6H_5)$.

Prepared by action of benzoyl chloride on the sodium salt of para-cresol. Insoluble in water. Soluble in ether and hot alcohol. Properties antiseptic.

BENZO-PHENONEID.

Non-irritant germicide similar to pyoctanin. Corneal ulcers, purulent keratitis, etc. Soluble in 100 parts of water.

BENZOSOL. $C_6H_4OCH_3OCOC_6H_5$.

Synonyms: Guaiacol Benzoate ; Benzoyl-Guaiacol.

This is a compound of guaiacol, in which a hydrogen atom of its hydroxyl is replaced by benzoyl. It is obtained by the action of benzoyl chloride on the sodium salt of guaiacol. Benzosol occurs as a colorless, crystalline powder, inodorous, tasteless, melting at 50° C. (122° F.); insoluble in water, readily soluble in alcohol. It contains 54% of guaiacol.

This compound is employed as an antiseptic in the treatment of phthisis, the dose being 0.25 to 0.5 Gm. (4 to 8 grains).

BENZOYL-ANILIDE. See Benzanilide.

BENZOYL-BETA-NAPHTHOL. See Benzonaphthol.

BENZOYL-EUGENOL. $(C_3H_5.C_6H_3(OCH_3)CO_2C_0H_5)$
Prepared by the action of benzoic acid on eugenol; it occurs in acicular crystals, melting at 70.5° C. (159° F.), free from odor and taste, insoluble in water, readily soluble in alcohol and ether. Recommended as an anti-tubercular.

BENZOYL-GUAIACOL. See Benzosol.

BENZOYL-PARA-CRESOL. See under Cresol.

BENZOYL-PSEUDO-TROPEIN HYDROCHLORIDE. See Tropa-Cocaine Hydrochloride.

BENZOYL-SULPHONIC-IMIDE. See Saccharin.

BENZOYLTROPEINE.
Silky needles forming soluble salts. Local anæsthetic.

BETA-NAPHTHOL-ANTIPYRINE. See Naphthopyrin.

BETA-NAPHTHOL-BENZOATE. See Benzonaphthol.

BETA-NAPHTHOL-CAMPHOR. See Naphthol-Camphor.

BETA-NAPHTHOL-CARBONATE. See Naphthol-Carbonate.

BETA-NAPHTHOL-DISULPHONATE OF ALUMINUM. See Alumnol.

BETA-NAPHTHOL-MONOSULPHONATE OF CALCIUM. See Asaprol.

BETA-PHENETIDYL-CROTONIC-ETHYL ESTER. See Phenetidyl.

BETA-PHENYL-ACRYLIC ACID. See Phenyl Acrylic Acid.

BETA-PHENYL-PROPIONIC ACID. See Phenyl Propionic Acid.

BETA-RESALGIN. See Resalgin.

BETOL. C_6H_4 (OH) COO.$C_{10}H_7$
Synonyms: Naphthalol; Naphthosalol; Salinaphthol; Salicylicnaphthylether.
Salol and betol are derivatives of salicylic acid, differing from one another in that in the former, a hydrogen atom of the salicylic acid is replaced by a phenyl group, (C_6H_5) while in the latter, it is replaced by a naphthyl group $(C_{10}H_7)$. Betol is prepared by heating together a mixture of beta-naphthol-sodium, sodium salicylate and phosphorus oxychloride. It forms a white, inodorous and tasteless, crystalline powder, which melts at 95°C. (203°F.), almost insoluble in water, but dissolves readily in alcohol and ether. Betol is employed as an intestinal antiseptic; it is also of value in treatment of vesical catarrh and articular rheumatism. The dose is 0.3 to 0.5 Gm. (5 to 8 grains).
ALPHOL is the corresponding salicylic ester of a-naphthol.
It is employed in doses of 0.5 to 1.0 Gm. (8 to 15 grains) in treatment of articular rheumatism and gonorrhœic affections.

BISMUTAL.
A mixture of bismuth sodium phosphate (Bismuth Phosphate) and sodium salicylate. Recommended as an antiseptic. When used as a dusting powder it is diluted with 5 parts of starch, or as a solution 1 to 4%; as an ointment, 10 to 20%.

BISMUTH COMPOUNDS.
The various organic salts of bismuth are prepared by interaction between a solution of bismuth nitrate [Bi (NO₃)₃] and a salt of the organic acid, the resulting bismuth compound precipitating. In order to obtain a clear aqueous solution of bismuth nitrate it is necessary either to dissolve the crystals, first in glycerin (2:3), then diluting with water, or to add sufficient nitric acid to the mixture of bismuth nitrate and water to effect a clear solution; in this latter instance it is necessary that the solution of organic salt be slightly alkaline. Bismuth nitrate may be also dissolved in a 25% sodium chloride solution or in acetic or diluted nitric acid. These basic bismuth compounds are decomposed by acids.

ALBUMINATE. A pale gray or white insoluble powder, containing about 9% of bismuth. Employed in gastric and intestinal cramps, in doses of 0.5 to 1 Gm. (8 to 15 grains).

BENZOATE (Sub-Benzoate.) Bi(C₆H₅CO₂)₃. Bi(OH)₃. By the heat of a water bath, 20 parts of bismuth nitrate Bi(NO₃)₃ are dissolved in 30 parts of glycerin, then diluted with 70 parts of water and poured slowly into a solution of 20 parts of sodium benzoate in 1,000 parts of water. (This order must not be reversed). The resulting precipitate is washed with warm water until the washings no longer react for nitric acid (Diphenylamine T. S), then dried at a temperature not above 80° C. Bismuth benzoate forms a white insoluble powder, which is employed as a dusting powder for torpid ulcers ; also used internally, being preferred by many to bismuth salicylate.

CERIUM-SALICYLATE forms an insoluble, pink-colored powder, being a valuable remedy in the treatment of diseases of the gastric and intestinal mucous membranes. Dose 1 to 2 Gm. (15 to 30 grains)

DITHIOSALICYLATE, or *THIOFORM*.   See under Dithiosalicylic Acid.

NAPHTHOLATE. Obtained by adding a solution of beta-naphthol in an alkali to a solution of bismuth nitrate, the latter being dissolved by aid of glycerin (see above) or a dilute acid. The precipitated bismuth naphtholate is well washed and dried at low temperature. It forms a brownish, tasteless powder, insoluble in water; it is employed as an intestinal antiseptic, in doses of 1 to 2 Gm. (15 to 30 grains.)

NAPHTHO-GLYCERITE. Recommended as a specific for gonorrhœa. Composition unknown.

OLEATE ($C_{17}H_{33}CO_2)_3Bi$. Obtained by precipitating a solution of sodium oleate with a solution of an equivalent amount of bismuth trinitrate; the solution of the latter being prepared as above directed. An insoluble powder possessing emollient and mild astringent properties; employed in various skin diseases.

OXYIODIDE, *SUBIODIDE*, or *BASIC IODIDE*. Bi OI. Crystalline bismuth trinitrate 95.4 Gm. is dissolved in 120 Cc. of glacial acetic acid; this solution is then poured, under constant stirring, into a solution of 33.2 Gm. of potassium iodide and 50 Gm. of sodium acetate in two liters of water. The precipitated oxyiodide is then washed by decantation, strained off and dried at 100° C. This is a reddish-brown, heavy powder, insoluble in all solvents, except acid and alkali solutions, by which it is decomposed. Recommended as an antiseptic, combining the action of iodine and bismuth.

OXYCHLORIDE (Subchloride). This is made by pouring a solution of bismuth trinitrate in diluted nitric acid into a solution of common salt. The white precipitate is well washed and then dried. Its medicinal uses are the same as those of the subnitrate.

OXY-IODO-GALLATE, or *AIROL.* C_6H_2 $(OH)_2$ CO_2BiOHI, is a substitute for iodoform, applied in ointment (water-free lard or lanolin) or dusted on in powder. It is light grayish-green in color; has no odor or taste.

PHENOLATE, or *CARBOLATE*, Bi $(OH)_2$. C_6H_5O. Prepared by interaction between a solution of bismuth trinitrate (see above) and a solution of sodium phenolate. Gray colored, insoluble, inodorous powder, used as an intestinal antiseptic in doses of 0.5 to 1 Gm. (8 to 15 grains) also externally as an antiseptic dusting-powder.

PHOSPHATE. (Soluble.) Obtained by fusing together bismuth oxide, caustic soda and phosphoric acid, pulverizing the resulting mass. This product contains 20% Bi_2O_3, and is very soluble in water; its solutions are rendered turbid by the addition of acids, alkalies, or by boiling. Recommended as an intestinal disinfectant, also in treatment of catarrh of the stomach, in doses of 0.2 to 0.5 Gm. (3 to 8 grains.)

PYROGALLATE (HELCOSOL). [$C_6H_2(OH)_2O]_2BiOH$. Prepared by dissolving 150 parts of pyrogallic acid in 650 parts of a 25% sodium chloride solution, and adding this solution to 316 parts of bismuth trichloride dissolved in 1000 parts of a salt solution of the same strength. After digesting on a water bath for one half hour the solution is poured into water, and the basic bismuth salt thus precipitated is washed with water acidulated with nitric acid until the washings are free from chlorides. Forms a yellow insoluble powder (60% Bi.), which is recommended as an internal and external antiseptic. Helcosol has a slightly different constitution, the formula being $C_6H_3(OH)_2OBi.(OH)_2$; it contains 56.6% of metallic bismuth and the powder is of a greenish-yellow color.

RESORCINATE. A solution of bismuth trinitrate is added to a solution of resorcin in excess of alkali. It forms a yellowish-brown powder which contains about 40 per cent. of Bi_2O_3. This compound is employed in treatment of chronic and acute catarrh of the stomach. The dose is not known.

SALICYLATE. (Bi $(C_7H_5O_3)_3$.Bi $(OH)_2+3H_2O$). The following process of L. Wolmann (Apoth. Ztg. 9.978) yields a bismuth salicylate of constant composition. Twenty-five parts of metallic bismuth in coarse powder is added in small portions to 125 parts of nitric acid (sp. gr. 1.20), heated to 75° to 90° C.; toward the end of the operation the temperature is increased to boiling. After standing several days, the fluid is decanted and evaporated to low bulk and crystallized. The crystals of bismuth nitrate are washed with a little water containing nitric acid, and, after draining, 48.6 parts of the crystals are dissolved in about 200 parts of dilute acetic acid and the solution rendered alkaline by the addition of aqua ammoniæ. The precipitate is well washed by decantation, until the wash water ceases to give a blue color on addition of a piece of zinc and a few drops of iodide of starch solution. The precipitate is brought to a paste by triturating with a little water in a mortar, then adding 13.8 parts of salicylic acid and heating on a water bath until the blue white color changes into that of a yellow-white. The mass is then collected on a muslin strainer, pressed and dried at a temperature not above 75°C. The bismuth salicylate thus obtained is a white, inodorous, tasteless and insoluble powder containing 64.65 per cent. of Bi_2O_3.

SUBGALLATE. See Dermatol.

SULPHITE. Prepared by interaction between solutions of sodium sulphite and bismuth nitrate, the latter being brought into solution by means of glycerin (see Bismuth Salts). Bismuth sulphite possesses an antiseptic and antifermentive action, being employed as such in intestinal disorders. Dose same as the subnitrate.

TRIBROMO-CARBOLATE. (Bismuth Tribromo phenate.) $Bi_2O_3(C_6H_2Br_3OH)$. A yellow, neutral, insoluble powder, inodorous, tasteless, containing 57 to 61% of Bi_2O_3. Used as an intestinal antiseptic in cholerine, cholera, inflammatory condition of the mucous intestinal membrane. Dose, 0.5 to 1 Gm. (8 to 15 grains.)

VALERIANATE. This is made by mixing 32 parts of bismuth subnitrate made into a thick paste with water, with a solution of sodium carbonate 11 parts, valeric acid 9 parts and water 30 parts; this mixture is allowed to digest for one hour, frequently stirring; the undissolved precipitate is collected, washed with cold water and dried at 30° C. It forms a white insoluble powder, possessing a strong valerian-like odor. Bismuth valerianate possesses the effect of the bismuth salts in addition to the anodyne action of the valeric acid. Dose is 0.05 to 0.25 Gm, (1 to 4 grains).

BOLDIN.

A principle obtained from the leaves of the *Boldoa chiliensis.* It is a white, amorphous, bitter powder, almost insoluble in water, readily soluble in alcohol, ether and chloroform. Employed as a tonic, also as hypnotic. Dose 0.064 Gm. (1 grain).

BORAL.

Synonym: Aluminum Boro-Tartrate.

Through the interaction between aqueous solutions of borax and aluminum sulphate, aluminum borate is obtained according to the following equation: $3Na_2B_4O_7+Al_2(SO_4)_3 = 3Na_2SO_4+Al_2(B_4O_7)_3$. The resulting precipitate is washed with water till free from sodium sulphate, then 1 part of this aluminum borate is dissolved by the heat of the water bath, in 10 parts of water, by means of ½ part of tartaric acid and evaporated to dryness at not above 40°C. The resulting aluminum-boro-tartrate (Boral) forms a soluble crystalline powder, which is recommended as an astringent antiseptic, either dry or in aqueous solution.

BOR-SALICYLATE.

A soluble and harmless antiseptic obtained by triturating together 32 parts of sodium salicylate and 25 parts of boric acid with a small amount of water; the mass soon becomes hard, when it is dried and powdered.

BOROCITRIC ACID.

This combination of boro and citric acids forms a white soluble powder, which is employed as a solvent for urates and phosphates in urinary calculi, gout, etc. Dose, 0.3 to 1.3 Gm. (5 to 20 grains.)

BOROPHENYLIC ACID. (Phenyl Boric Acid.) $C_6H_5B(OH)_2$.

Obtained by the action of phosphorus oxychloride on a mixture of boric acid and phenol. Forms a soluble white powder of mild aromatic taste; melts at 201° C. (400° F.) Employed as a preservative agent (1:5000 sol.)

BORO-SALICYLATE OF GLYCERIN.

Boric and salicylic acids, when heated in the presence of glycerin, dissolve in large proportions, but, on cooling, the mixture soon becomes turbid, forming a thick and granular mass. If this mixture be heated anew until it boils and a small quantity of calcined magnesia be then added, the solution after cooling remains perfectly limpid. The product thus obtained is miscible with water in all proportions. This boro-salicylate of glycerine enables the operator to obtain extemporaneously a solution containing equal parts of the two acids at a degree of concentration impossible with any other method. Moreover, the microbicide and antiseptic properties of the salicylic and boric acids are in nowise affected by their being transformed into a neutral or basic salt. The following is the formula:

Boric acid	10 Gm.
Salicylic acid	10 "
Distilled water	10 "
Thirty per cent. dist. glycerin	40 "

Heat the mass in a flask until it boils and then add 1 Gm. calcined magnesia; reduce the fire and evaporate all the water, obtaining, after cooling, 50 Cc. of the glycerole or boro-salicylate, 5 Cc. of which will contain exactly 1 Gm. each of salicylic and boric acids.

BOROSALYLIC ACID. $BOH(OC_6H_4 COOH)_2$.

A combination of boric and salicylic acids in molecular proportions. Recommended as an antiseptic instead of salicylic acid.

BOROSOL:

A colorless liquid of acid reaction, containing, according to various analyses, aluminum tartrate, boric acid, salicylic acid and free tartaric acid in aqueous solution. Borosol is recommended as a wash for perspiring feet.

BRASSICON.

A new headache remedy, a green-colored mixture, consisting, according to the *Süddeutsche Apotheker Zeitung*, of 2 Gm. oil of peppermint, 6 Gm. camphor, 4 Gm. ether, 12 Gm. alcohol and 6 drops of mustard oil.

BROMAL-HYDRATE. C Br. $COH + H_2O$.

Synonym: Tri-Brom-Aldehyde-Hydrate.

A mixture of alcohol 1 part and bromine 4 parts is heated to 140°C., then allowed to cool slowly; on standing, crystals of bromalhydrate separate. It forms colorless crystals, soluble in water, melting at 53.5° C. (128.3° F.); when heated to 100° C. it is decomposed into bromine and water.

Bromal-hydrate is employed as a sedative and antispasmodic, its action being the same as that of chloral hydrate, being given, however, in smaller doses, [0.1 to 1 Gm. (1½ to 15 grains)].

BROMALIN. ($C_6H_{12}N_4$, C_2H_5Br).

Synonym: Hexamethylene-Tetramin-Brom-Ethylate.

This compound appears in colorless scales, or as a white crystalline powder, readily soluble in water. It is administered to women and children in doses of ½ to 4 Gm. (30 to 60 grains), as a nervine and sedative.

BROMAMIDE. $C_6H_2Br_3.NH_2.HBr$.

Synonym: Tri-Brom-Anilin Hydrobromide.

Nitrotribrombenzol is reduced by means of nascent hydrogen, the resulting product being treated with hydrobromic acid. This occurs in colorless, tasteless crystals, melting at 117°C. (242.6°F.), being employed as an antineuralgic in doses of 0.6 Gm. (10 grains).

BROMHÆMOL. See under Hæmol.

BROMOFORM. C H Br₃.

Synonym: Tri-Brom-Methane.

This analogue of chloroform is prepared by the action of sodium hypobromite on acetone. It forms a clear, colorless liquid, of chloroformic odor and taste; its specific gravity is 2.9 and boiling-point 152°C. (305.6°F.). Bromoform is only very slightly soluble in water, but readily in alcohol. It is employed in treatment of whooping-cough in daily doses of 5 to 20 drops.

BROMOL. $C_6H_2Br_3OH$.

Synonyms: Tri-Brom-Phenol; Bromphenol.

This compound is obtained by pouring an aqueous solution of bromine in an aqueous solution of phenol, a white crystalline precipitate resulting. The precipitate is washed and crystallized from alcohol. Bromol forms colorless crystals, which are insoluble in water, very soluble in alcohol, fatty and volatile oils. It is employed externally as an antiseptic in solution (1:30 olive oil), or ointment (1:10), or as a dusting powder, in the treatment of diphtheria it is used in a 4% glycerin solution.

BROMO-PHENOL.

Synonym: Ortho-Bromo-Phenol.

This is a dull, violet-colored liquid, having a phenol-like odor; obtained by treating phenol with bromine. It is employed in the form of a 1 to 2% ointment in the treatment of erysipelas.

BROM-PHENOL. See Bromol.

BURSIC ACID.

The active principle of the *Bursa pastoris*. Forms a pale, yellow, hygroscopic mass, of astringent taste. Employed subcutaneously as a hæmostatic, being of equal value to ergot.

BUTYL-CHLORAL-HYDRATE.. C Cl₃-CH₂-CH₂-COH-H₂O.

Synonym: Croton Chloral.

A current of chlorine gas is passed through paraldehyde until saturated, the resulting butyl-chloral is purified by distillation, and brought in contact with water. Butyl-chloral-hydrate forms colorless, crystalline scales, which melt at 78°C., soluble in 30 parts of cold water, readily soluble in alcohol and ether.

It is employed as a hypnotic, in doses of 1 to 1.5 Gm. (15 to 24 grains).

BUTYL-HYPNAL. See under Antipyrine.

BUTYROMEL.

A mixture of 2 parts of fresh unsalted butter and 1 part of honey; intended as a substitute for cod-liver-oil.

BUXINE. $C_{16}H_{21}NO_3$.

An alkaloid from the bark of *Buxus sempervirens*, identical with berberine. Recommended as a tonic and febrifuge; in doses of 1 to 2 Gm. (15 to 30 grains).

CADMIUM SALICYLATE. $(C_6H_4 (OH) CO_2)_2Cd$.

This salt is prepared by the action of salicylic acid upon cadmium hydrate or carbonate, or by precipitating barium salicylate with cadmium sulphate. When prepared by the first method, molecular quantities of the two substances are heated together with water until solution takes place, then evaporated to low bulk and crystallized. Thus obtained cadmium salicylate forms fine, white, tabular crystals, which melt at 300°C. (572°F.) and dissolve in 24 parts of water at 100°C., and 68 parts at 23°C.; also soluble in alcohol, ether and glycerin. The latter method of preparation yields an amorphous powder. Cadmium salicylate possesses a more energetic antiseptic action than the other salts of cadmium, being recommended in treatment of purulent ophthalmia.

CÆSIUM BITARTRATE. $Cs_2C_4H_4O_6$.

Forms colorless prismatic, strongly refractive crystals which are readily soluble in water. This and the corresponding Rubidium compound were recommended by Schaefer in nervous heart palpitation, in doses of 0.18 to 0.3 Gm.

CAFFEINE SALTS (DOUBLE SALTS AND DERIVATIVES). $C_8H_{10}N_4O_2A+xH_2O$.

Among the large number of salts of caffeine that have been introduced, a very few have received attention, among the very important of these are the *carbolate*, *phthalate* and *boro-citrate* which are readily soluble, the former two being recommended for hypodermic use.

CAFFEINE CHLORAL, $C_8H_{10}N_4O_2 \cdot C Cl_3COH$. This is a combination of chloral and caffeine in molecular proportions. It is crystalline and soluble in water. Caffeine-chloral is a sedative and analgesic. Dose 0.2 to 0.3 Gm, (3 to 5 grains).

CAFFEINE-SODIUM BENZOATE. Is prepared by evaporating an aqueous solution of one part of caffeine in one of sodium benzoate dissolved in 3 parts of water. It forms white crystalline crusts. Owing to the ready solubility of this and the following double salts, they are especially suitable for subcutaneous use.

CAFFEINE-SODIUM SALICYLATE and *CAFFEINE-SODIUM-CINNAMATE* are prepared in the same manner as the above benzoate, employing equal parts of caffeine and the respective organic sodium salts.

CAFFEINE-SODIUM SULPHONATE. See Symphorol.

CAFFEINE TRI-IODIDE. (Di-Iodo–Caffeine–Hydriodide.) $(C_8H_{10}N_4O_2I_2HI)_2+3H_2O$. This is prepared by adding a solution of hydriodic acid to a weak alcoholic solution of caffeine. It forms dark-green prisms which are readily soluble in alcohol. Internally it acts like a weak preparation of iodine, the dose being 0.12 to 0.24 Gm. (2 to 4 grains).

ETHOXY-CAFFEINE. $C_8H_9N_4O_2(OC_2H_5)$. Is prepared by boiling monobrom-caffeine with caustic potash. It forms crystalline needles which are less soluble in water than caffeine, melting at 138°C. (280.4°F.) Ethoxy-caffeine has a similar action to caffeine, being also narcotic. Dose is about 0.2 Gm. (3 grains).

IODO-CAFFEINE. Is prepared by passing sulphuretted hydrogen into a solution of potassium iodide and caffeine. Iodo-caffeine forms colorless crystals soluble in water; unstable, decomposed by hot water; it is employed in cardiac affections in doses of about 0.3 Gm. (5 grains.)

IODOL-CAFFEINE. $C_8H_{10}N_4O_2 C_4I_4NH$. A crystalline compound prepared by the interaction of molecular quantities of iodol and caffeine. It is a gray color, inodorous, tasteless and practically insoluble in the ordinary solvents. Iodol-caffeine is employed as an antiseptic like iodol, of which it contains 75%.

IODO-THEOBROMINE. Is prepared in like manner to the above caffeine compound. It is used as a diuretic, alterative, and in cardiac affections. Dose, 0.3 to 0.5 Gm. (5 to 8 grains.)

CAHINCIN. See Caincic Acid.

CAINCIC ACID. $C_{40}H_{64}O_{16}$.

Synonyms: Cahincic Acid; Cahincin.

A crystalline principle obtained from the root of the *Chiococca anguifuga* (cainca root). Forms crystalline flakes, inodorous, of a bitter, astringent taste, soluble in water and alcohol. Given in doses of 2 to 4 grains (0.13 to 0.25 Gm.) as a diuretic and cathartic; as an emetic, in doses of 8 to 15 grains. (0.5 to 1 Gm.)

CAJUPUTOL.   See Eucalyptol.

CALCIUM BORATE.

Obtained by interaction between aqueous solutions of borax and calcium chloride. Recommended as an antiseptic dusting-powder in treatment of moist eczema, burns, etc.; likewise internally in doses of 0.2 to 0.4 Gm. (3 to 6 grains) for diarrhœa of children.

CALCIUM PERMANGANATE.

Deliquescent brown needles, decomposing readily when in contact with organic substances.   It is claimed to be a more powerful bactericide than corrosive sublimate.

CALCIUM SALICYLATE. $\left[C_6H_4 < ^{OH}_{COO}\right]_2 Ca + 2H_2O$

Salicylic acid is neutralized with an equivalent amount of calcium carbonate (free from iron) in the presence of hot water, the filtered solution is then evaporated and crystallized. Calcium salicylate forms a white, crystalline powder, inodorous and tasteless, almost insoluble in cold water.

It is employed in the diarrhœa of children in gastro-enteritis, the dose being 0.5 to 1.5 Gm. (8 to 24 grains).

CALCIUM SULPHOPHENATE (SULPHOCARBOLATE).

Prepared by neutralizing sulphocarbolic acid with calcium carbonate.   It forms a white, almost inodorous, stable, astringent, bitter powder, freely soluble in water and alcohol.   Recommended because of its strong antiseptic, disinfectant and astringent properties, given internally in 1% aqueous solution.

CALOLACTOSE.

An intestinal disinfectant, said to consist of a mixture of calomel (1), bismuth subnitrate (1), and lactose (8).

CAMPHOID.

This is prepared by dissolving pyroxylon, 1 part, in a solution of 20 parts of camphor in alcohol. It constitutes a thick, colorless fluid, which, because of its rapidity in drying, leaving a thin film when applied to the skin, serves as an excellent vehicle for iodoform, chrysarobin, etc.

CAMPHOPYRAZOLON. $C_{17}H_{20}N_2O$.

This is a compound of phenyl-hydrazine and campho-carboxylic acid. It occurs in fine crystalline needles, melting at 132° C. (269.6° F.), insoluble in water and ether, soluble in alcohol. Campho-pyrazolon is proposed as a substitute for camphor.

CAMPHOR, NAPHTHOL.   See Beta-Naphthol Camphor.

CAMPHOR, PHENYLATED.   See Phenol Camphor.

CAMPHOR SALICYLATE.

This is prepared by fusing together 84p. of camphor and 65p. of salicylic acid, which solidifies to a crystalline mass. It may be obtained in well formed crystals by recrystallization from benzol. Soluble 1 in 20 in the fatty oils; almost insoluble in water and glycerin. Employed externally in ointment form as an application in lupus and various skin diseases, internally in treatment of certain diarrhœal complaints.   Dose 0.05 to 0.25 Gm. (4-5 to 3.8 grains).

CAMPHOR, SALOL.   See under Salol.

CAMPHORIC ACID. $C_8H_{14}(COOH)_2$.

This is a dibasic acid obtained by the action of hot nitric acid on camphor. It forms white, scaly crystals, odorless, melting at 186.5° C (368°F.)   It is soluble in 200 parts of cold and 10 parts of boiling water; very soluble in alcohol.

It is employed in treatment of night sweats of consumptives, likewise in acute and chronic diseases of the respiratory tract. The dose is from 1 to 1.5 Gm. (15 to 24 grains.) When applied topically it is used in a solution of from 1 to 4% strength.

CAMPHORONIC ACID. $C_8H_{11}(COOH)_3$.

Synonym: Iso-propyl carbylic Acid.

Obtained by the oxidation of campholic acid. It forms soluble white needles, hygroscopic ; melts at 136° C, (276.8° F.) Recommended as an antiseptic.

CANNABIN.

An alkaloid isolated from *Cannabis Sativa*, or Indian Hemp. Cannabin forms a brown, syrupy liquid, which is employed as a hypnotic, the dose being 0.06 to 0.3 Gm. (1 to 5 grains).

CANNABIN TANNATE forms a yellowish-gray colored powder of bitter and slightly astringent taste, only slightly soluble in water, alcohol and ether, very soluble in acidulated water. It is employed as a hypnotic in nervous sleeplessness, the dose being 0.25 to 1 Gm. (3 to 15 grains).

CANNABINDON.

A red syrupy fluid, obtained from *Cannabis indica*, soluble in alcohol and ether, but insoluble in water. Recommended as hypnotic in hysteria and insanity in doses of 0.032 to 0.097 Gm. (½ to 1½ grains) once daily. Women half this dose.

CANNABINON.

A resinous body obtained from the flowering tops of the *Cannabis Sativa*. Cannabinon appears in form of a dark-brown, soft resin, insoluble in water; soluble in alcohol, ether, chloroform, fatty and volatile oils. Recommended as a sedative and hypnotic, in doses of 0.03 to 0.1 Gm. (½ to 1½ grains).

CANNONIN.

A proprietary disinfectant.

CANTHARIDIN. $C_{10}H_{12}O_4$.

The active vesicating principle obtained from the *Cantharis vesicatoria* and other members of the family of *Coleoptera*. Cantharidin forms colorless crystals, which are insoluble in water, but very soluble in chloroform; it dissolves quite readily in ether and the fatty oils, with caustic alkalies it forms salts soluble in water. This principle is frequently employed in place of cantharides. The salts, or cantharidates, are employed hypodermically in treatment of tuberculosis, 0.2 Gm. of cantharidin and 0.4 Gm. of potassium hydrate being dissolved in 1.000 Cc. of distilled water, of this solution 0.2 to 0.4 Cc. (0.0001 to 0.0002 Gm.) being employed for a subcutaneous injection.

CAPITCURA.

A proprietary antipyretic and analgesic.

CARDINE.

A clear yellow fluid, prepared by digesting the finely chopped hearts of sheep with an equal quantity of glycerin and boric acid solution in a well-closed vessel, and subsequently filtering. Employed subcutaneously in 3 to 5 Cc. doses as a heart tonic and diuretic.

CARDOL.

A blistering oil obtained from the pericarps of the *Anacardium occidentale*, by extraction with ether. Employed externally as a vesicant.

CARNIFERRIN.

This is a meat preparation, of German production, being a combination of the phosphoric acid of the body with 30% of iron. It is given in 3 to 5 grain doses for children and 8 grains for adults. It is readily absorbed and tasteless, and mixes well with acid or alkaline solutions.

CARNOLIN.

An aqueous solution containing 1.5% of formaldehyde; specific gravity 1.0035. Recommended as a harmless disinfectant and preservative for food.

CARPAIN. $C_{14}H_{27}NO_2$.

An alkaloid obtained from the leaves of *Carica papaya*. Carpain forms handsome colorless crystals of a bitter taste, melting at 121° C. (249.8° F.); it readily unites with acids, forming crystalline salts.

This alkaloid is employed as a substitute for digitalis, being given in doses, hypodermically, of 0.006 to 0.01 Gm. (1-10 to 1-6 grain).

CARVACROL. $C_{13}H_{13}OH$.

A phenol found in the essential oil of *Origanum species*. It forms a thick fluid, which boils at 235°C. Carvacrol possesses powerful antiseptic properties.

CARVACROL IODIDE. $C_{19}H_{13}OI$.

This is prepared analogous to aristol, by the action of iodine upon an alkaline solution of carvacrol. It constitutes a brown-colored powder, which becomes soft at 50°C. (122°F.), melting at 90°C. (194°F.) to a brown fluid. It is insoluble in water, slightly soluble in alcohol, readily in ether, chloroform and olive oil.
Carvacrol iodide is employed as a substitute for iodoform.

CASCARINE.

A principle isolated from the *Cascara sagrada*; it is also identical with *rhamno-xanthine*, which occurs in the *Rhamnus frangula*. This principle occurs in colorless, tasteless needles, melting at 200° C. (392° F.), insoluble in water, but soluble in alcohol. Cascarine is recommended in doses of 2 to 3 grains (0.15 to 0.2 Gm.) daily in treatment of habitual constipation.

CASEIN-SODIUM.

A soluble, chemically pure albuminoid body, which may be dissolved in milk, cocoa, bouillon, etc., without in the least interfering with the taste of these. Ten grammes of this preparation represent 500 Gm. of milk. Casein-sodium is obtained by evaporating and drying the calculated amounts of casein and sodium hydrate in vacuo.

CATECHU-TANNIC ACID.

A reddish-brown powder obtained from the *Acacia catechu*. It is employed as an astringent to check diarrhœa, hæmorrhage; also bleeding gums, ulcerated nipples, epistaxis, etc.

CETRARIN. $C_{18}H_{16}O_8$.

A bitter principle obtained from Iceland moss, (*Cetraria Islandica*). It forms colorless crystals of a bitter taste, difficultly soluble in cold, but very soluble in hot alcohol.
Cetrarin increases peristalsis, likewise the secretion of saliva, bile and pancreatic juice. Internal dose is 0.1 to 0.2 Gm. (1.5 to 3 grains).

CHEMIA.

A proprietary antiseptic.

CHINASEPTOL. See under Chinoline.

CHININUM BIMURIATICUM CARBAMIDATUM. See Quinine Dihydrochloride-
[Carbamate.

CHINOL.

Synonyms: Chinoline or Quinoline Monohypochloride.

White, crystalline, odorless powder, almost insoluble in water. **Soluble in alcohol.**
Antipyretic. Dose, 3 to 5 grains.

CHINOLINE. C_9H_7N.

A tertiary amine, obtained by the distillation of quinine or cinchonine with potassium hydrate, or, as synthesized by Skraup, by heating a mixture of nitrobenzol, aniline, glycerin and sulphuric acid. Pure chinoline is a yellowish-colored liquid, of aromatic odor, its specific gravity being 1,084; boiling at 237°C. (458.6°F.). It is almost insoluble in water, very soluble in alcohol and ether. Chinoline is antiseptic, antizymotic and antipyretic; being employed chiefly as a tooth- and mouth-wash (0.2%).

DERIVATIVES.

Chinoline unites readily with the acids forming soluble crystalline salts.

CHINOLINE SALICYLATE. $C_9H_7N \cdot C_7H_6O_3$. A white crystalline powder, soluble in 80 parts of water, very soluble in alcohol, ether, glycerin and the oils. Antifebrile and antiseptic in doses of 0.5 to 1 Gm. (8 to 15 grains).

CHINOLINE TARTRATE. $(C_9H_7N)_3 (C_4H_6O_6)_4$. Occurs in colorless, rhombic crystals, soluble in 80 parts of cold water, less so in hot water, soluble in 150 parts of alcohol. Its properties and doses are similar to the above.

ACETO-ORTHO-AMIDO-CHINOLINE C_9H_8N (NHCH$_3$CO). This preparation is an analogue of acetanilid, in which chinolin replaces phenyl. It forms colorless crystals, which melt at 102.5°C. It possesses antipyretic properties.

DIAPHTHERIN (OXY-CHIN-ASEPTOL), HO.C$_9$H$_6$N-HSO$_3$.C$_9$H$_4$OH-C$_9$H$_6$N.OH. This is a compound of one molecule of oxychinolin with one molecule of phenolsulphonate of oxychinolin. It forms clear, yellow crystals, soluble in water and melting at 85°C. (185°F.). Diaphtherin possesses antiseptic properties equal to those of carbolic acid, as also the advantages of solubility and of being non-poisonous. It is employed in ½ to 1% solution. The solution readily attacks surgical instruments.

DIAPHTOL (CHINASEPTOL). C_9H_5 (OH) (SO_3H) N. This is an ortho-oxychino-line-meta-sulphonic acid, bearing the same relation to chinoline as phenol-sulphonic acid does to benzol. It forms yellowish-colored crystals which are only slightly soluble in cold water, melting at 295°C. (563° F.). Its aqueous solution, like that of diaph-therin, gives a green color with ferric chloride. The properties of diaphtol are similar to those of diaphtherin

KAIRIN. $C_9H_{10}(C_2H_5)$ NO.HCl. *Synonyms:* Ethyl Kairin ; Kairin A ; Oxy-Chino line-Ethylhydride. This is a derivative of chinoline; its method of preparation is complicated. Kairin was the first synthetical substitute for quinine. It was recommended as an antipyretic in doses of 0.5 to 1 Gm. (8 to 15 grains).

KAIROLIN. $C_9H_{10}(C_2H_5)$ N.H_2SO_4. *Synonyms:* Kairolin A and M ; Chinolin-ethylhydride (A): Chinolinmethylhydride (M). Kairin M. is the hydrochloride of x-oxy-chinolin-methyltetrahydride, while kairolin A and M are the acid sulphates of eth' l-chinoline-tetrahydride and methyl-chinoline-tetrahydride, respectively. These remedies are not employed since the discovery of other antipyretics.

LORETIN This is a meta-iodo-ortho-oxychinoline-ana-sulphonic acid, (C_9H_4I-OH-SO_3H-N) a powerful antiseptic discovered by Claus. It forms a yellow, inodorous, crystalline powder, which is only very slightly soluble in water (1:1000), insoluble in ether and the ۰ ils, melting at about 270°C. (518°F.). Loretin forms a valuable substitute for iodoform, having the advantage of being free from odor and toxic effect. It is employed as a dusting powder, either alone or diluted ; in 5 to 10% ointments and 0.1 to 0.2% aqueous solutions.

RHODANATE. (Sulphocyanate.) $C_9H_7N < ^H_{SCN}$ Forms white crystals, melting at 140° C. (284° F.), soluble in cold water to the extent of 3.5%, very soluble in hot water. A powerful bactericide in 0.3 to 3% solutions.

SULPHOCYANATE. See Chinolin Rhodanate.

CHLORALAMID. CCl_3 CH.OH. NH.COH.

Synonym: Chloral-Formamide.

This is prepared bv interaction between chloral (CCl_3COH) and formamide ($HCONH_2$). Chloralamid forms colorless crystals, which melt at 115° C. (239° F.), soluble in 20 parts of cold water and about 1½ of alcohol ; it should not be heated with water.

It is employed as a substitute for chloral hydrate, in doses of 1 to 3 Gm. (15 to 45 grains).

CHLORAL-AMMONIUM. $CCl_3CH.OH.NH_2$.

This should not be confused with *Chloralamid.* Chloral-ammonium is obtained by passing a current of dry ammonia gas into a solution of chloral in chloroform. It forms colorless needles, which melt at 34° C. (183.2° F.), almost insoluble in water ; when boiled with water it is decomposed into chloroform and ammonium formate.

It is used as a hypnotic and analgesic, in doses of from 1 to 2 Gm. (15 to 30 grains).

CHLORAL-ANTIPYRINE (Monochlor-Antipyrine). See under Antipyrine.

CHLORAL-CAFFEINE.

A combination of chloral and caffeine in aqueous or alcoholic solution, prepared by a patented process, possessing (over caffeine) the advantage of ready solubility.

CHLORAL FORMAMIDE. See Chloralamid.

CHLORAL-HYDROCYANATE.

Synonyms: Chloral Hydrocyanin, Chloral Cyanhydrin.

White rhombic plates with odor of hydrocyanic acid and chloral. Soluble in water, alcohol and ether. Fairly stable in solution. Decomposed by alkalis. 1.29 parts dissolved in 9 parts distilled water equivalent to U. S. P. 2 per cent. hydrocyanic acid. 1 to 180 — bitter almond water of Germ. Pharm. 646 parts hydrocyanate contain 1 part pure HCN.

CHLORALIMID. CCl_3CH-NH.

This body is obtained by heating chloral ammonium. Chloralimid is a crystalline powder, which is almost insoluble in water, easily in alcohol ; mineral acids decompose it into chloroform and ammonia salt. Its properties are those of a hypnotic, in doses of 1 to 4 Gm. (15 to 60 grains.) Decomposed by mineral acids.

CHLORALOSE. $C_8H_{11}Cl_3O_6$.

Synonym: Anhydrogluco-Chloral.

This is a compound of chloral with grape sugar. Chloralose forms fine colorless needles, which melt at 184° to 185° C. (363.2° to 366.8° F.), soluble in 170 parts of cold water, readily so in alcohol. It is employed as a hypnotic (substitute for chloral), in doses of 0.2 to 0.5 Gm. (3 to 8 grains).

CHLORALOXIMES.

These are a class of bodies which consist of compounds of chloral with various oximes. Among the more important of these are *chloralacetoxime, chloral camphoroxime, chloral acetaldoxime, chloral benzaldoxime, chloral nitroso-beta-naphthol,* etc. These compounds are soluble in alcohol and decomposed by heating with water. They are intended as hypnotics, the dosage having not been determined.

CHLORAL URETHANE. See Uralium.

CHLOR-METHYL. See Methyl Chloride.

CHLOROBROM.

A mixture of equal parts of chloralimid and **potassium bromide.** Used as a hypnotic, especially in the treatment of the insane.

CHLOROIODOLIPOL.

A chlorine substitution product of phenol, creosote and gualacol, **recommended** for inhalation in treatment of chronic diseases of the air passages.

CHLOROLIN.

An antiseptic solution containing chiefly mono- and tri-chlor-phenol. It is recommended as an effectual disinfectant for cesspools, closets, hospitals, etc. As an anti septic wash in surgical operations a 2 to 3% solution is strong enough. An antiseptic soap is also prepared from it.

CHLORO-PHENOL. (Tri-Chlor-Phenol). $C_6H_2 < {}^{Cl_3}_{OH}$

This is a derivative of carbolic acid, in which three atoms of hydrogen are replaced by chlorine; it occurs in the form of colorless, needle-like crystals, with an odor of phenol. This is employed as a 1 to 2% ointment in treatment of erysipelas.

CHLOROSALOL.

The chlorphenylic ether of salicylic acid. Has been recommended as a surgical antiseptic by Prof. Girardi, of Berne. Further experiments will be made as to its availability for internal administration.

CHLORPHENOL. $C_6H_4 < {}^{Cl\ (1)}_{OH\ (2)}$

Synonyms: Monochlorphenol; Ortho-Mono-Chlor-Phenol.

The preparation employed under this name consists of a mixture of ortho-monochlorphenol (7 pt.) and alcohol, eugenol and menthol (together 3 pt.) This liquid is employed in diseases of the respiratory organs, from 16 to 80 drops being inhaled daily.

PARA-MONOCHLORPHENOL. $C_6H_4 < {}^{Cl\ (3).}_{OH\ (2).}$

This is a crystalline body resulting from the chlorination of phenol, possessing greater antiseptic power than the other two isomers. It melts at 37°C. (98.6°F.), is readily soluble in alcohol, but sparingly in water; it is employed as a 1 to 2% ointment in treatment of erysipelas.

CHLORSALOL. $C_6H_4 (OH) CO.OC_6H_4Cl$.

Synonyms: Chlorphenol Salicylate; Salicylic-Chlorphenol-Ester.

This is prepared by the action of phosphorus penta-chloride on a mixture of ortho- and para-chlorphenol. Chlorsalol, that is the ortho-phenyl-ester, forms colorless crystals which melt at 55°C. (131°F.), while those of the para-phenyl-ester melt at 72°C. (161.6°F.). Both are insoluble in water and soluble in alcohol. Chlorsalol is employed as a substitute for salol, being more energetic in its action.

CHLORYL.

This name has been applied to a mixture of methyl and ethyl chlorides. It is a liquid at 0°C.; employed as an anæsthetic, being milder in effect than ethyl chloride.

CHROATOL.

Greenish-yellow aromatic crystals, insoluble in water, slightly in ether and chloroform. Quite soluble in alcohol, benzol and acetic ether. Dermal application in psoriasis, alopecia, etc., in powder or ointment.

CHROMOSOT.

Said to consist chiefly of sodium sulphite and sulphate. Used as a disinfectant.

CHRYSAROBIN. $C_{30}H_{26}O_7$.

A principle obtained from Goa powder, which is a concretion found in the stem and branches of the *Andira Araroba*. It is a light yellow, crystalline powder, very slightly soluble in water, slightly soluble in alcohol, ether and chloroform, freely soluble in alkalies. By oxidation, chrysarobin is converted into chrysophanic acid. It is employed chiefly in treatment of various skin diseases, in ointment of 10% strength.

CINCHONIN-HERAPATHIT. See Antiseptol.

CINCHONIN-IODOSULPHATE. See Antiseptol.

CINEOL. See Eucalyptol.

CINNAMIC ACID. $C_6H_6CH=CH.COOH$.

Synonyms: Acid Cinnamylic ; Beta-Phenylacrylic Acid.

This occurs naturally in Peru and Tolu balsams ; it is obtained synthetically by heating benzaldehyde and acetyl-chlorid together under pressure. It forms colorless to yellowish, glossy plates, of melting point 133°C. (271.4°F.). Insoluble in cold, but quite soluble in boiling water.

It is employed, in form of an emulsion or alcoholic solution, as hypodermic injection in the treatment of tubercular affections.

℞ Acid cinnamylic................... 5.0
    Ol. Amygdal. dulc.................. 10.0
    Vitelli Ovi... ....No.1.
    Solut. Natrii chlorati (0.7%)....... q. s.
    Misce ut fiat emulsio.
    D. S. Injection 0.1 to 1 cc.

CINNAMYL-EUGENOL. See under Eugenol.

CINNAMYL-GUAIACOL. See under Guaiacol.

CITROPHEN. $C_3H_4OH\left(\begin{smallmatrix}CONH\\OC_2H_5\end{smallmatrix}>C_6H_4\right)_3$.

A compound of citric acid and para-phenetidin $(C_6H_4 < \begin{smallmatrix}OC_2H_5\\NH_2\end{smallmatrix})$. It forms a white crystalline powder of acidulous taste, soluble in about 40 parts of cold water (which recommends it for subcutaneous injections); melts at 181° C. (358° F.) Citrophen is recommended as an antipyretic and antineuralgic in doses of 0.5 to 1 Gm. (8 to 15 grains.)

COCAINE SALTS. $C_{17}H_{21}NO_4A$.

Only the more important of the new combinations will be given. The doses are essentially the same as in the hydrochlorate.

COCAINE-ALUMINUM SULPHATE. This double compound of cocaine and aluminum sulphate is obtained by mixing together solutions of cocaine and aluminum sulphate, evaporating and crystallizing. Nothing definite is known as to its application and dosage.

COCAINE BORATE. Is employed for subcutaneous injections and eye douches. It is preferable to all other salts of cocaine because of the stability of its aqueous solutions and the indifference of the boric acid.

COCAINE LACTATE. Is a soft mass, readily soluble in water ; it is of value in treatment of cystitis.

COCAINE NITRATE. Is employed in combination with silver nitrate in treatment of diseases of the genito-urinary tract.

COCAINE PHENATE or *CARBOLATE,* Forms a soft mass, insoluble in water, very soluble in alcohol. This salt is employed subcutaneously as a local anæsthetic, also as a local application ; stronger solutions are required to produce the same degree of anæsthesia as with the hydrochlorate.

COCAINE SACCHARATE. Forms hydroscopic, crystalline plates. A 5% solution of this salt corresponds to a 4% solution of the hydrochlorate. Because of its sweet taste it is preferred for throat applications.

COCAPYRIN.

A mixture of 100 parts of antipyrin and 1 part of cocaine. Administered in pastilles each containing 0.2 Gm. (3 grains.) of antipyrin and 0.002 Gm. (1-32 grain.) of cocaine.

CODEINE HYDRIODATE. See Iodic Acid.

CODEINE PHOSPHATE. $C_{18}H_{21}NO_3.H_3PO_4+2H_2O$.

This salt occurs in fine colorless needles, a bitter taste, is readily soluble in water and sparingly in alcohol. Codeine phosphate is adapted as a substitute for morphine for administration to children, also employed in most affections of the respiratory organs, etc. Dose 0.025 to 0.05 Gm. (1-3 to ¾ grains).

CODOL. See Retinol.

COLCHICEIN. $C_{21}H_{23}(OH)NO_6+\frac{1}{2}$ Aq.

When colchicin is boiled with dilute sulphuric acid it is converted into colchicein and methyl alcohol. Colchicein forms white crystals, which are soluble in water and alcohol. It should be dispensed with caution, the dose being 0.001 Gm. (1-64 grain) subcutaneously.

COLCHICINE. $C_{22}H_{25}NO_6$.

An alkaloid obtained from the *Colchicum autumnale*. Colchicine forms a yellowish white amorphous powder which is readily soluble in water and alcohol; melts at 145° C. (293° F.) It is employed in treatment of rheumatism, sciatica, etc., in doses of 1-120 to 1-160 grain.

CONIINE HYDROBROMIDE. $C_8H_{17}N.HBr$.

This is the hydrobromate of the liquid alkaloid Coniine, which is obtained from the seeds of the *Conium maculatum*. It forms colorless crystals, which are soluble in water. Employed in the treatment of cardiac asthma in doses of 0.003 to 0.005 Gm. (1-20 to 1-12 grain).

CONVALLAMARIN. $C_{23}H_{44}O_{12}$.

A glucoside obtained from roots of the *Convallaria majalis*. It is a white powder, very bitter, soluble in water and alcohol. Employed as a cardiac stimulant in doses of 0.05 Gm. (¾ grain).

CONVALLARIN.

Glucoside from *Convallaria majalis*, not to be confounded with Convallamarin. Crystals very soluble in alcohol, insoluble in water. Drastic purgative.

CONVOLVULIN.

A glucoside obtained from the root of *Ipomœa purga* and other plants of the same genus. An amorphous mass, insoluble in water, readily so in alcohol and acetic acid. Convolvulin is a powerful purgative; dose 0.06 to 0.13 Gm. (1 to 2 grains).

CORNUTIN.

One of the active principles of ergot. A reddish to yellowish-colored powder which readily forms salts with acids. It is recommended to relieve hæmorrhage arising from abortion, also to increase the vigor of labor pains. The dose is 0.005 Gm. (1-12 grain).

CORONILLIN.

A glucoside obtained from the seeds of *Coronilla scorpioides*; forms a pale yellow, bitter powder, soluble in water and alcohol, almost insoluble in ether. Strengthens the action of the pulse and increases diuresis. Dose 0.06 to 0.13 Gm. (1 to 2 grains).

COTOIN. $C_{22}H_{18}O_6$.

A neutral principle obtained from the coto bark. It forms an amorphous or crystalline powder, which melts at 130°C. (266° F.), slightly soluble in water, freely soluble in alcohol and ether. Cotoin is employed in treatment of cholera, it is also said to relieve night sweats. The dose is 0.03 to 0.3 Gm. (½ to 5 grains).

COUMARIN. See Cumarin.

CREATIN.

Constituent of muscular tissue. Opaque, white solid with bitter, acrid taste. The monohydrate occurs in transparent prisms. Soluble in 70 water. Muscular and digestive tonic. Dose, 1½ grains, 3 to 5 times daily.

CREOLIN. See under Cresol.

CREOSAL.

A compound obtained by heating a mixture of equal parts of tannin and creosote at 80° C., then adding phosphorus oxychloride and continuing the heat until no more gas is evolved. The mass is mixed with dilute sodium hydrate solution, whereby creosal separates; this is well washed and dried. Creosal is a hygroscopic, dark-brown powder, which is readily soluble in water alcohol and glycerin. Creosal is recommended in powder form or aqueous solution in treatment of inflammation of the air passages.

CREOSOL. $C_6H_3CH_3$ (OH) $(O.CH_3)$.

Synonyms: Homo-Pyrocatechin-Metyl Ether; Homoguaiacol.

This occurs, along with guaiacol, as a constituent of beech-wood tar creosote. It is an oily-like liquid, of aromatic odor, boiling at 220° C. (428° F.), only slightly soluble in water. Recommended as an antiseptic.

CREOSOTE-CALCIUM-CHLORHYDRO-PHOSPHATE.

A white, syrupy mass, consisting of a mixture of creosote carbonate and calcium-chlorhydro-phosphate. Recommended in phthisis and scrofula. Dose 5 to 10 grains.

℞ Creosote-calc.-chlorhyd.-phosp............5 to 10 Gm.
  Mucilag. chondri.......................... 15 "
  Ol. Amyg. dulc........................... 25 "
  Syr. Tolutani............................ 25 "
  Aq. flor. aurant......................... 75 "
  M. f. emulsio.
Dose two teaspoonfuls daily.

CREOSOTE CARBONATE. See under Guaiacol.

CREOSOTIC ACID. $C_6H_3 <^{CH_3}_{OH}$ COOH

Synonyms: Cresotinic Acid; Oxytoluic Acid; Homosalicylic Acid.

This may exist as an ortho, meta, or para modification, hence is frequently designated in the plural, as creosotic acids. These bear the same relation to toluene ($C_6H_5CH_3$) that salicylic acid bears to benzene (C_6H_6), being then hydroxytoluic acid. They are prepared from the sodium cresylates by a process (Kolbe's) analogous to that used in the manufacture of salicylic acid. The para compound, which crystallizes in white needles melting at 151°C. (303.8°F.), is the only one that is employed in medicine, in the form of a sodium salt.

CREOSOTOL. See under Guaiacol.

CRESALOLS. C_6H_4 (OH) $CO_2C_6H_4.CH_3$.

Synonyms: Cresol Salicylates; Cresol Salols.

Ortho-, meta-, and para-cresalol are the salicylic esters of the cresols, analogous to betol and salol, and prepared in a similar manner. A mixture of sodium salicylate and cresylate, in molecular proportions, is heated with phosphorus oxychloride; either ortho, meta, or para cresalol is obtained according to the sodium salt used. These three isomeric cresalols form bulky, white, crystalline powders, insoluble in water, soluble in alcohol and ether, and sparingly so in oils. The ortho-cresalol melts at 35° C. (95° F.), the meta-cresalol at 74° C. (165.2° F.), and the para-cresalol at 39° C. (102.2° F.)

Externally the cresalols are recommended as antiseptic dusting-powders, also internally as a substitute for salol, being split up in the system into cresol and salicylic acid.

CRESOL SALICYLATE. See Cresalols.

CRESOL SALOLS. See Cresalols.

CRESOL, or THE CRESOLS. $C_6H_4 <^{OCH_3 (1).}_{OH \ (3).}$

Synonyms: Cresylic Acid; Meta-Cresol; Kresol; **Meta-Phenol.**

The Cresols, of which three isomerides exist (ortho-meta-para) are homologues of phenol and derivatives of toluene. They are obtained by the fractional distillation of that portion of coal-tar oil which comes over between 190°—210° C. The three isomerides are exceedingly difficult to separate; of these the meta-cresol is the most powerful antiseptic. All the cresols possess a creosote-like odor, their antiseptic properties are superior to those of carbolic acid and they are far less poisonous. The only hindrance to their general employment is their insolubility in water. During the past few years various soluble preparations introduced have again brought them into notice; in these, the cresol (o-m-p) is rendered soluble by the addition of soap (*Sapocarbol, Lysol, Creolin, Phenolin, Sapocresol,* etc.), or an alkali forming soluble cresylates (*Solveol, Solutol*) or by conversion into soluble sulphonic acid derivatives (*Artmann's Creolin*).

PREPARATIONS.

The following are preparations of Cresol which have been introduced as antiseptics in surgery.

ANTINONNIN is a preparation of ortho-dinitro-cresol potassium, appearing in trade in the form of a soap. It is employed in a 1 part to 1000 solution for the destruction of insects and fungi.

BENZOYL-PARA-CRESOL, (Para-Cresol Benzoate) C₆H₅COO C₆H₄ (CH₃), is prepared by the reaction of phosphorus oxychloride on a mixture of benzoic acid and para-cresol. It forms a crystalline compound which melts at 70° C. (158° F.), insoluble in water, readily soluble in alcohol and ether. It is used as an antiseptic.

CREOLIN. This is said to be an emulsion of cresol obtained by means of resin soap. It consists of a brownish-black, syrupy liquid, which, when mixed with water, forms a more or less turbid mixture; with alcohol, ether or chloroform, it forms a clear solution. Creolin is employed in the pure condition, as a 1 to 2% solution; in ointments, dusting-powder or dressings, 10%; internally, it is administered in capsules containing 5 minims.

CRESOL IODIDE. Very light, yellow powder, with disagreeable odor, readily soluble in alcohol, ether, chloroform and oils, insoluble in water. Adheres to hands, instruments, etc., like resin. Antiseptic, allaying inflammation in nasal diseases.

IZAL. According to Squibb, this consists of an emulsion containing about 30% of a new oil produced by a patent process employed in the manufacture of a special form of coke. It is claimed that its antiseptic power is greater than that of carbolic acid, while it is practically non-poisonous.

KRESIN. This is a clear brown liquid containing 25% of cresol and 25% of cres-oxy-acetate. It is miscible in all proportions with water.

LYSOL. The fraction of coal-tar oil which boils between 190° to 200° C. is dissolved in fat and subsequently saponified, with the addition of alcohol. It forms a clear brown syrupy liquid, containing 5 % of the cresols; it is miscible with water, forming a clear, saponaceous frothing liquid. With all other solvents it is miscible in all proportions. Experiments have shown lysol to be five times stronger than carbolic acid in antiseptic power. The strength of the solution employed is usually 0.3, 1 or 2%.

PARACRESOL is a patented disinfectant, which is said to mix with water in every proportion, yielding a neutral and almost odorless solution.

PHENOSALYL. This is a solution of carbolic, salicylic and benzoic acids, (which have been fused together), in lactic acid. It forms a thick syrupy liquid, which is soluble in cold water to the extent of 7%, readily in warm water, also in alcohol and ether. In antiseptic power it is superior to carbolic acid, being at the same time less toxic. It is employed in 1 to 2% aqueous solutions.

PIXOL and *RESOL* are solutions of wood tar in soap.

SANATOL is a solution of crude cresol in sulphuric acid.

SAPOCARBOL, KRESAPOL and *PHENOLIN* are solutions of crude cresols in soap (potassium).

SAPROL is a dark brown, oily substance, consisting of a mixture of the crude cresols in an excess of hydrocarbons, obtained from the refining of petroleum. A drawback to its use is its inflammability, also that it does not mix with water.

SOLUTOL. This is an alkaline solution of sodium-cresol in an excess of cresol. It is not suited for surgical dressings or like uses, because of its caustic alkalinity. It is a valuable disinfectant for use in the household and hospital, effectually disinfecting water closets, sinks, cess-pools, etc.

SOLVEOL is a solution of cresol in sodium cresotate; it forms a useful disinfectant analogous to Creolin, Lysol, Saprol and Solutol. It is less caustic than solutol, possessing the advantage over creolin and lysol of not exhibiting the greasiness characteristic of these. It is a dark colored liquid, nearly odorless, of a neutral reaction and miscible with water in all proportions. It is especially applicable for surgical uses, a ½% solution being employed in dressings and a one to twelve for spray apparatus. It is claimed that a ½% solution is more active than a 3% solution of carbolic acid.

TRICRESOL is a concentrated preparation of the three cresols (ortho,meta,para), free from all impurities. It forms a clear colorless liquid, of specific gravity 1.045, and soluble to the extent of 2½% in water. A 1% solution of tricresol corresponds to a 3% solution of carbolic acid, having therefore three times the disinfectant value of the latter.

CRESYLIC ACID. See Cresol.

CROTON CHLORAL. See Butyl-Chloral-Hydrate.

CROTON-CHLORAL-HYDRATE. See Butyl-Chloral-Hydrate.

CRYOSTASE.

A mixture of equal parts of carbolic acid, camphor, saponin and traces of oil of turpentine. Becomes solid when heated, and liquid when cooled to below 0°C. Recommended as an antiseptic.

CRYSTALLOSE.

A name applied to a very soluble crystalline sodium salt of pure saccharin. In consequence of the presence of crystal water, the sweetening power, compared with that of cane sugar, is reduced to 400 times the latter instead of 500 times, as is the case in the pure amorphous saccharin.

CUBEBIC ACID. $C_{28}H_{30}O_7$.

A principle obtained from the cubeb by extraction with caustic alkalies and afterwards liberated on addition of an acid. Cubebic acid forms a waxy-like body, readily soluble in alcohol and ether, becoming brown on exposure to the air. According to Bernatzik it possesses the antiblennorrhagic properties of cubebs; dose 0.3 to 1 Gm. (4 to 15 grains).

CUMARIN.

Synonym: Coumarin.

The crystallizable, odorous constituent of the *Tonca bean*, also prepared synthetically. It melts at 67° C. (152.6° F.), only slightly soluble in water, readily in alcohol and ether. Cumarin is employed for the purpose of masking the odor of medicinal agents such as iodoform, etc.

CUPRATIN.

A copper albuminoid preparation, analogous to Ferratin.

CUPROHÆMOL. See under Hæmol.

CURARINE. $(C_{18}H_{35}N)$.

This alkaloid, which belongs to the strychnos family, forms a yellow amorphous hygroscopic powder, soluble in water and alcohol. The pure alkaloid is given in doses of 0.25 to 0.7 milligrammes (1-250 to 1-90 grain) to relieve attacks of tetany. Hypodermically 1 Cc. (16 M.) or less of a solution of 25 Ctg. (3¾ grains) of curarine, in 5 Gm. (1¼ fld. dr.) each of glycerin and distilled water.

CUTAL.

Synonym: Aluminum Boro-Tannico-Tartrate.

Obtained by pouring a mixture of five parts of an aqueous tannin solution (1:4) and 80 parts of an aqueous borax solution (1:10) into a solution of 3 parts of aluminum sulphate in 12 parts of water, stirring constantly. The resulting precipitate is filtered off, washed, spread on glass plates and dried at low temperature. This preparation is insoluble in water; hence, in order to render it soluble, 1 part of it is dissolved in 10 parts of water by means of 1.2 parts of tartaric acid; the solution evaporated to dryness at a low temperature yields a soluble aluminum boro-tannico-tartrate or cutal. This is recommended as an astringent antiseptic, either in dry form or in aqueous solution.

CUTIN.

A name applied to a soft material which is intended as a substitute for silk or catgut, and which may also be employed to prevent gauze from adhering to wounds. It is prepared from the gut of cattle by carefully removing the serous and mucous membranes from the muscular layer, and digesting the latter in a 2% solution of pepsin. It is then treated with trioxybenzoic acid and hydrogen peroxide, which harden the membrane. Thus prepared cutin is very soft, adheres smoothly to the wound, is capable of being absorbed, and may be sterilized by dry heat.

CYSTINE. $C_{11}H_{14}N_2O$.

An alkaloid obtained from the *Cytisus laburnum*. Forms an inodorous, deliquescent, white, crystalline mass, of bitter taste, readily soluble in alcohol and water, insoluble in ether; chiefly employed as a nitrate. Cystine, as regards its physiological action, stands between strychnia and curare, being used in paralytic migraine; also in cardiac diseases.

Dose 0.003 to 0.005 Gm. (1-21 to 1-12 grain).

CYTISINE. $C_{11}H_{14}N_2O$.

An alkaloid obtained from the *Cytisus laburnum*, which forms white deliquescent crystals, soluble in water and alcohol, melt at 155° C. The hydrochlorate of cytisine is used as a nervine, given in paralytic migraine, whooping cough, asthma, in doses of 0.003 to 0.005 Gm. (1-20 to 1-12 grain) subcutaneously.

DATURINE.   See Hyoscyamine.

DELPHININ.

An alkaloid obtained from the seed of *Delphinium staphisagria.* Forms small crystals of a bitter taste, insoluble in water and soluble in alcohol and ether. Exerts a powerful action on the heart, like aconitin; employed in treatment of spasmodic asthma, dropsical affections and neuralgia, in doses of 0.01 to 0.02 Gm. (1-6 to 1-3 grain).

DERMATIN.

A new skin protecting preparation used in dermatology and consisting of from 5 to 7 parts of salicylic acid, 7 to 15 parts of starch, 25 to 50 parts of talc, 30 to 60 parts of silicic acid, and 3 to 9 parts of kaolin, according to the strength desired.

DERMATOL. $C_6H_2(OH)_3COOBi(OH)_2$.

Synonyms: Subgallate of Bismuth; Basic Gallate of Bismuth.

To a solution of 15 parts of crystallized bismuth trinitrate in 30 parts of glacial acetic acid diluted with about 200 parts of water, is added with constant stirring a warm solution of 5 parts of gallic acid in 250 parts of water. The yellow precipitate is washed until free from nitric acid, then dried on porous plates, or the following method may be employed; Take 57-2 Gm. of bismuth subnitrate (assaying 81.4% of Bi_2O_3), dissolve with the aid of heat in 71 Gm. of commercial nitric acid (sp. gr. 1-36), and dilute with 12 Ccm. of water. After cooling this solution it is gradually diluted with 75 Ccm. of water, which does not cause precipitation. The solution is filtered and precipitated by pouring into it under constant stirring a solution consisting of 37.8 Gm. (theory requires 37.6 Gm.) of gallic acid in 800 Ccm. of water. The gallic acid is dissolved with the aid of heat, but the solution is cooled to about 30° C. before being used. The resulting precipitate is washed by decantation until the washings are but faintly acid. It is then collected on a filter and washed with the aid of suction until the washings no longer give a perceptible reaction with diphenylamin T. S.   The subgallate is then dried at about 55° C. and sifted, when a soft, bright yellow powder results. It is entirely soluble in sodium hydroxide solution, and therefore free from subnitrate. When treated with diphenylamin T. S. according to Fischer's method it does not give a reaction for nitric acid.   Dermatol forms a bright yellow, isodorous and tasteless powder, insoluble in the usual solvents. It is an excellent dry antiseptic in all varieties of surgical practice. Internally in doses of 0.25 to 0.5 Gm. (4 to 8 grains) in treatment of diarrhœa.

DERMOL.

Synonym: Bismuth Chrysophanate.

This is described by Trojescu as an amorphous yellow-colored powder, insoluble in all of the usual solvents, consisting of a mixture of chrysarobin and bismuth hydroxide. It is probably intended to be used as an antiseptic application in various skin diseases.

DESOXY-ALIZARIN.   See Anthrarobin.

DEXTRO-SACCHARIN.

This consists of a mixture of saccharin 1 part and glucose 2000 parts.

DIABETIN. $C_6H_{12}O_6$.

Synonyms: Lævulose ; Fruit Sugar.

Fruit Sugar (*Fructose*) is found in most sweet fruits, together with an equal amount of grape sugar ; it is formed, together with grape sugar, in the so-called *inversion* or decomposition, of cane sugar by boiling with acids ; the mixture of the two is called *invert sugar.* This diabetin is prepared from invert sugar by mixing the latter with calcium-hydroxide, the liquid lime compound of dextrose is removed and the residual solid is the lime compound of lævulose ; this latter calcium salt is decomposed with carbonic acid, liberating the lævulose. This is a colorless, odorless, crystalline powder, readily soluble in water and alcohol. Diabetin is recommended as a sweetening agent for diabetic patients.

DIACETANILID. $C_6H_5N(C_2H_3O_2)_2$.

The old method of preparation of this salt is by heating together phenyl mustard oil and acetic acid.   It is now prepared by heating acetanilid with glacial acetic acid at 200° to 250° C.   The reactions product is taken up by hot petroleum ether and crystallized, while the unaltered acetanilid remains behind undissolved.   The physiological action of this compound is similar to, but stronger, than that of acetanilid.

DIACETYL TANNIN.   See Tannigen.

DIAMINE. (N_2H_4)

Synonym: Hydrazine.

Occurs in colorless crystals, analogous to hydroxylamine hydrochloride.   Powerful reducing agent.   A general poison to animal and vegetable life, destroying germs, bacteria, etc.

DIAPHTHERIN. See under Chinolin.

DIAPHTHOL. See under Chinolin.

DI-BROMO-GALLIC ACID. See Gallobromol.

DI-CHLORAL-ANTIPYRINE. See under Antipyrine.

DI-CHLORACETIC ACID. $CHCl_2COOH$.

This is obtained by the action of chlorine gas on glacial acetic acid. It forms a colorless, pungent fluid, of boiling-point 190° C. (374° F.). Is employed as a cauterizing agent.

DI-CHLOR-METHANE. See Methylene Chloride.

DIETHYL ACETAL. See Acetal.

DI-ETHYL-SULPHON-DI-ETHYL-METHANE. See Tetronal.

DI-ETHYL-SULPHON-DI-METHYL-METHANE. See Sulphonal.

DI-ETHYL-SULPHON-METHYL-ETHYL-METHANE. See Trional.

DI-ETHYLENE-DIAMINE. See Piperazine.

DIGITALEIN. (Schmiedeberg.)

A glucosidal principle obtained from the leaves of the *Digitalis purpurea*. It constitutes a yellowish, amorphous powder which is soluble in water and alcohol. The properties and dose of this glucoside are similar to those of digitalin.

DIGITALIN. (Digitalinum.)

A glucoside obtained from the leaves of the *Digitalis purpurea*.

DIGITALINUM CRYSTALLISATUM NATIVELLE. (French.) Forms fine crystalline needles, which are almost insoluble in water and soluble in alcohol. It has been recommended in treatment of inflammation of the lungs, and also feebleness of the heart's action. Dose 0.00065 to 0.001 Gm. (1-100 to 1-64 grain.)

DIGITALINUM VERUM KILIANI. (German.) $(C_6H_8O_2)x$. Occurs as a white amorphous powder, soluble in 1000 parts of water and in 100 parts of dilute alcohol. It melts at 217°C. (422.6°F.). This digitalin "verum" exerts the characteristic effects of digitalis leaves (Bohm and Pfaff), being administered in doses of 0.00025 Gm. (1-300 grain).

DIGITOXIN. $C_{31}H_{52}O_7$.

According to Schmiedeberg, digitoxin is the most active of the several glucosides which constitute commercial "digitalin," being essentially identical with "Nativelle's digitalin." Digitoxin forms white crystalline needles, which are insoluble in water, the dose being 0.00033 to 0.00065 Gm. (1-200 to 1-100 grain).

DI-HYDRO-RESORCIN.

This is prepared by the action of sodium amalgam on a solution of resorcin in boiling water, carbonic acid gas being passed through the solution during the reaction. The unconverted resorcin is removed by shaking the solution with ether, then after acidulating, the solution is shaken a second time with ether, which extracts the dihydroresorcin, which remains as a syrup-like liquid on the evaporation of the ether. It soon congeals to a solid mass on standing, which, by reerystallization, yields glossy prisms, which melt at 104° to 106° C. (219.2° to 222.8°F.), very soluble in water, alcohol and chloroform. Di-hydro-resorcin is recommended as an antiseptic.

DI-IODO-CARBAZOL.

This is prepared by heating a solution of carbazol to the boiling point and adding iodine. It occurs in yellow, odorless laminæ, which melt at 184° C. (363.2° F.), insoluble in water, but readily soluble in alcohol, chloroform, etc. Recommended as an antiseptic.

DI-IODOFORM. C_2I_2.

Synonym: Ethylene periodide,

This is obtained by the action of iodine on a solution of acetylene iodide in carbon disulphide. It forms bright yellow, inodorous, crystalline needles, which are insoluble in water, sparingly soluble in alcohol, and readily so in chloroform; melting at 192° C. (377.6° F.). Diiodoform contains 95.5% of iodine, and is recommended as a substitute for iodoform. Exposure to light causes its decomposition, hence it should be kept in a dark place.

DI-IODO-SALICYLIC ACID. $C_6H_2I_2(OH)COOH$.

This is obtained by the action of iodine and iodic acid on salicylic acid. It forms a white crystalline powder, of sweet taste, melting between 220° to 230° C, (428° to 446° F.), almost insoluble in cold water, slightly soluble in hot water and very soluble in alcohol and ether.

It is employed as an analgesic and antipyretic ; dose 1.5 to 4. Gm. (24 to 60 grains) per day.

DI-IODO-SALOL. See under Salol.

DIMETHYL-ACETAL. $CH_3-CH(OCH_3)_2$.

Synonym; Æthyliden-Di-Methyl-Ether.

This is obtained by the oxidation of a mixture of ethyl and methyl alcohols. It forms a colorless, ethereal liquid of sp. gr. 0.867, boiling at 64° C. (147.2° F.) Dimethyl-acetal is employed as an anæsthetic, either alone or mixed with half its volume of chloroform.

DI-METHYL-ETHYL-CARBINOL. See Amylene Hydrate.

DI-METHYL-KETONE. See Acetone.

DI-METHYL-PIPERAZINE TARTRATE. See under Piperazine.

DIOXY-ANTHRANOL. See Anthrarobin.

DIPHTHERIA ANTITOXINE.

Behring's curative serum is the blood-serum of **animals immunized by the injection of** the diphtheria toxine.

All infectious diseases like diphtheria are produced by bacteria, which secrete very poisonous substances called "toxines" that eventually cause death. By the use of certain agents the action of these toxines may be counteracted, rendering the organism insensible to their poisonous effect. This insensibility or immunity may be acquired by a gradual habituation to a given poison. To Prof. Behring is due the credit of discovering that during this process of habituation an antitoxine is produced in the blood, which, when isolated and injected into the blood of a patient, effects a change in the susceptibility of the living organs to the action of the poison (toxines). The antitoxine is prepared as follows: A colony of diphtheria bacilli, after being placed in a suitable medium and under favorable conditions, multiply with great rapidity, secreting at the same time their poison or toxines. After a few weeks, when sufficient of the toxines has formed, the bacilli are destroyed by means of carbolic acid and by filtering through porous plates of clay the dead bacilli are removed from the solution of toxines. Of this solution, small amounts are injected into the blood of a healthy horse, producing a mild attack of the disease ; this procedure is then repeated for several months, the doses of toxine being steadily increased until the animal becomes habituated to the poison. Then a quantity of blood is withdrawn from the animal, and the serum or aqueous portion is separated from the red blood corpuscles ; this serum constituting a light yellow liquid which contains the antitoxine of diphtheria. This serum is standardized by determining the quantity required for injection to neutralize a fatal dose of diphtheritic poison in a guinea pig ; the ratio between the quantity of antitoxine and the body weight of the animal furnishes a means of indicating in definite units the strength of the solution.

DIPHTHERIA-ANTITOXINE SOLUTION.

BEHRING'S. No. 1 equals 600 immunizing units; No. 2, 1,000; No. 3, 1,500. Hyp. inj. one-fourth of a vial. No. 1 as prophylactic, regular treatment full contents of either strength, according to case. In ½ oz. vials, varying measure but full unit value.

GIBIER'S. (N. Y. Pasteur Institute.) Identical with *Roux's.* Immunizing power 1:100,000 i. e., ½ ccm. prophylactic up to 110 lbs.; regular treatment 5 to 15 ccm. per day.

ROUX'S. Same description as *Gibier's.*

SCHERING-ARONSON. This preparation is supplied in vials containing 5 Cc., equivalent to 500 antitoxic normal units. The contents of a single vial are injected subcutaneously in mild cases, or at the onset of severer cases. The quantity contained in two vials (10 Cc.) may be used in cases that present very severe symptoms from the first, or in cases where this method of treatment has not from the first been carried out.

In malignant cases, particularly where there are laryngeal symptoms, three or four doses may be used (15 to 20 Cc.), depending on the age of the patient. For purposes of immunization the injection of 1 Cc. for small children and 2 Cc. for grown children and adults will be found sufficient. This preparation is a very permanent one, being rendered so by the addition of 4% Tricresol.

DIPHTHERICIDE.

Pastilles containing thymol, sodium-benzoate and saccharin. **Used as prophylactic** against diphtheria.

DISINFECTIN.

This is made by treating 5 parts of "masut" (the residue of naphtha-distillation) with 1 part of concentrated sulphuric acid; the resulting sulphonated product is then treated with 5 parts of 10% soda solution. This forms a brown liquid which, when diluted with water, is used as a disinfectant.

DISINFECTOL.

A mixture of hydrocarbons and crude cresols rendered soluble by the addition of alkali. It is a dark-brown liquid which gives a milky-like solution with water. It is employed as a disinfectant diluted with water.

DISPERMINE. See Piperazine.

DITHION. See under Dithiosalicylic Acids.

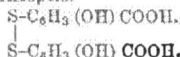

DI-THIO-CHLOR-SALICYLIC-ACID. $(S_2C_6H.Cl.OH.COOH)$

Obtained by heating a mixture of 27.6 parts of salicylic acid and 55 parts of sulphur chloride to 120° C., finally raising to 140° C. **It forms a reddish-yellow powder** which is recommended as an antiseptic.

DITHIOSALICYLIC ACIDS.
$$\begin{array}{l} S\text{-}C_6H_3\,(OH)\,COOH. \\ | \\ S\text{-}C_6H_3\,(OH)\,COOH. \end{array}$$

Of these acids nine isomers are possible, but only two of them have been introduced in medicine in the form of sodium and lithium salts, being distinguished as No. 1 and No. 2.

Salicylic acid and sulphuryl chloride, in molecular proportions, are heated together at 150° C., the resulting resinous like mass is dissolved in a solution of soda to which in turn a solution of sodium chloride is added, resulting in the precipitation of sodium dithiosalicylate No. 1, while the sodium salt No. 2 remains in solution. The acids are liberated from their corresponding sodium salts by the addition of hydrochloric acid.

SODIUM DITHIOSALICYLATE No. 1. Forms a yellowish, amorphous, soluble powder, which is employed as an antiseptic in veterinary practice, either as a wash (3 to 5%) or mixed with talcum or starch (5 to 50%) as a dusting powder.

SODIUM DITHIOSALICYLATE No. 2. Forms a gray, amorphous, hygroscopic and soluble powder, which is employed internally in treatment of muscular rheumatism and rheumatic fever; in antiseptic activity it is superior to sodium salicylate. Dose is 0.2 to 1 Gm. (3 to 15 grains).

DITHION. Is a mixture of the two sodium salts of dithiosalicylic acid. It is employed as an antiseptic wash (5 to 10%) and dusting-powder in veterinary practice.

THIOFORM. A basic bismuth salt of dithiosalicylic acid, introduced as a substitute for iodoform. It is prepared by adding a solution of sodium dithiosalicylate (1 or 2) to a solution of bismuth trinitrate, the latter salt being first dissolved in a little glycerin before diluting with water. The resulting precipitate, after washing and drying, constitutes a voluminous, yellow, insoluble, inodorous powder.

DI-THYMOL-IODIDE. See Aristol.
DI-THYMOL-TRIIODIDE. See Annidalin.
DIURETIN. $C_7H_7N_4O_2Na+C_6H_4\,(OH)\,COONa.$

Synonym: Sodio-Theobromine Salicylate.

The active constituent of this body is theobromine, an alkaloid which is closely related to caffeine. Diuretin is prepared by mixing aqueous solutions of sodio-theobromine and sodium salicylate, in molecular proportions, and evaporating to dryness. It constitutes a white amorphous powder, very soluble in water, decomposed by acid solutions. It is employed as a diuretic in doses of 0.5 to 1.5 Gm. (8 to 23 grains) for children, and 1.5 to 3 Gm. (23 to 45 grains) for adults.

UROPHERIN (*Lithium-Diuretin*). Is analogous to diuretin, being the corresponding lithio-theobromine-salicylate. It is a white powder, soluble in 5 parts of water, being employed as a diuretic in doses of 1 Gm. (15.5 grains).

DUBOISINE. $C_{17}H_{23}NO_3.$

This alkaloid, obtained from the leaves of *Duboisia myoporoides* is, according to Ladenburg, chemically identical with hyoscin; according to others identical with hyoscyamin. It is employed as a mydriatic, acting more rapidly and being less irritating than atropine; also employed as a hypnotic and sedative in treatment of hysteria and epilepsy. The *sulphate* occurs in hygroscopic crystals; being used in aqueous solution (0.065 Gm. to 30 Cc.) for the eye. The dose of the alkaloid is 0.0008 to 0.002 Gm. (1-80 to 1-30 grain).

DULCIN. $CO(NH_2)NH . C_6H_4OC_2H_5$.
Synonyms : Sucrol ; Valzin ; P-Phenetolcarbamid.

By the action of para-phenetidin on phosgene, dissolved in toluol, phenetidin-carbon-oxychloride is formed ; this product on treatment with ammonia yields dulcin. This sweetening agent forms colorless crystals, which melt at 173° C (343 4° F.) soluble in 800 parts of water at 15° C. (59° F.), and 55 parts of water at 100° C. (212° F.) ; it is soluble in 25 parts of alcohol, also in ether. Its sweetening power is about 200 times that of sugar. Dulcin is employed as a sweetening agent for the food of diabetics.

EMBELIC ACID. $C_9H_{14}O_3$.

A crystalline principle obtained from the fruit of the *Embelia Ribes*. It forms orange crystalline scales, inodorous, tasteless, soluble in alcohol and water, melts at 140° C. (284° F.) After neutralizing with ammonia and mixing with honey it is given in doses of 0.2 to 0.4 Gm. (3 to 6 grains.) For the expulsion of tape worm.

EMOL.

A kind of earth containing steatite with traces of lime and iron oxide, appearing as a soft, pink, impalpable powder. Recommended to be applied as a paste to remove epidermal masses, as well as horny epidermis in certain cases of eczema of the palm and sole.

ENTEROL.

A mixture of three isomeric creosols (pure) in the same proportions as they occur, as physiological products, in the intestinal tract. Enterol is recommended as an intestinal antiseptic. It possesses an unpleasant odor, and is usually administered in pills or capsules. A solution of 0.02 Gm. in 100 Gm. of water is administered in doses of 1 to 5 Gm. (15 to 75 grains) per day. Usually administered with a laxative.

EPHEDRINE. $C_{10}H_{15}ON$.

An alkaloid obtained from the leaves of the *Ephedra vulgaris*. The hydrochloride of this base, which forms colorless soluble crystals, is recommended as a mydriatic to replace homatropine. Instillation of 1 or 2 drops of a 10% solution into the eye produces marked dilation of the pupil, which lasts from 5 to 20 hours ; for daily application 2 to 3 drops of a 1% solution is sufficient.

ERGOTININE.   (Tanret).

This is a feeble basic principle obtained from ergot, of which, according to Tanret, it is the active principle. It occurs in colorless needles, which, on exposure to light, rapidly darken. According to Kobert, this principle, when pure, is inert.

ERYTHROPHLEINE.

This is an alkaloid obtained from the Sassy bark (*Erythrophleum guincense*). The hydrochloride of this base forms colorless crystals which are soluble in water. This alkaloid has been recommended as a local anæsthetic, internally as a cardiac stimulant. It greatly increases blood pressure, acting in a manner very similar to that of digitalis and of strophanthus. Subcutaneously, the lethal dose of erythrophleine hydrochlorate upon cats is about 0.003 Gm. (1-20 grain), 0.01 Gm. (1-6 grain) killing them within 15 minutes. It is intensely poisonous.

ESERIDINE. $C_{15}H_{23}N_3O_3$.

This is one of the alkaloids which occur in the calabar bean, along with physostigmine (eserine). Eseridine forms colorless crystals, melting at 132° C. (269.6° F.) Its physiological action is similar to that of eserine, being, however, six times weaker in effect. It has been recommended as a purgative in veterinary practice.

ETHOXY-CAFFEINE.   See under Caffeine.

ETHYL BROMIDE. C_2H_5Br.

Synonyms: Æther bromatus; Brom-ethyl; Mono-brom-ethane.

Ethyl bromide is obtained by the distillation of a mixture of ethyl-alcohol, sulphuric acid and potassium bromide. It forms a colorless, limpid, inflammable liquid, of a sweet chloroformic odor. It boils between 38° and 40° C. (100.4°—104° F.); its specific gravity lies between 1.445 and 1.450. It is not miscible with water, but freely with alcohol, ether, chloroform and the oils.

This preparation should be cautiously preserved, for the action of light and air causes its decomposition into bromine and hydrobromic acid, which is evident by its brown color, acid reaction and pungent odor; in this condition it should not be dispensed.

Ethyl bromide is employed as an anæsthetic in minor surgical operations. Narcosis is produced in from ½ to 1 minute, but lasts only a few minutes, unless fresh quantities are administered.

ETHYL CARBAMATE.   See Urethane.

ETHYL CHLORIDE. C_2H_5Cl.

Synonyms: Chlor-ethyl; Mono-chlor-ethane; Chelen.

Ethyl chloride is produced by the action of dry hydrochloric acid gas on absolute alcohol. At ordinary temperature it constitutes a gas, which is readily condensed to a liquid which boils at 10^0 C. (50^0 F.) Because of the intense cold (about -35^0 C.) produced by its evaporation, it is employed as a local anæsthetic. This liquid appears in commerce in small hermetically sealed tubes, terminated by a capillary point. When used, the point of the capillary is broken off, and the tube held in the hand, the warmth of which is sufficient to expel the liquid through the small orifice in a stream; this is directed to the surface where it is desired to produce local anæsthesia.

Because of its great inflammability, operations should not be performed **in the** proximity of a gas flame.

ETHYL IODIDE. C_2H_5I.

Synonyms : Æther Iodatus; Ethyl Iodide; Mono-iod-ethane.

Ethyl iodide is prepared by the action of iodine on a well-cooled mixture of amorphous phosphorus and absolute alcohol. This forms a colorless liquid which boils at 71^0 C. (160^0 F.); its specific gravity being 1.97.

Employed as an inhalation in treatment of bronchitis and dyspnœa.

ETHYL KAIRIN. See Kairin.

ETHYL NATRIUM. See under Sodium Salts.

ETHYL PHENACETINE. See under Phenacetine.

ETHYL URETHANE. See Urethane.

ETHYLENE BROMIDE. $C_2H_4Br_2$.

Synonyms: Æthylenum Bromatum; Brom-ethylene.

Ethylene bromide is prepared by passing ethylene gas through bromine. It forms a colorless, highly refractive liquid of chloroformic odor; its boiling point is 131.5^0 C. (269^0 F.) and specific gravity 2.170. It is insoluble in water, but miscible with alcohol and the fatty oils.

This ethylene bromide should not be confused with ethyl bromide, as the former produces marked toxic effects when inhaled.

Ethylene bromide is employed as an anti-epilepticum, **in doses of 0.1 to 0.3 Gm.** (1.5 to 5 grains) or 6 to 12 minims, in form of an emulsion.

ETHYLENE CHLORIDE. $C_2H_4Cl_2$.

Synonyms: Æthylenum Chloratum ; Æthylen Chlorid; Chloræthylene; Elaylum Chloratum; Liquor Hollandicus.

Ethylene chloride is produced by the action of chlorine on ethylene. This consists of a colorless ethereal liquid, of boiling point of 85^0 C. (185^0 F.) and specific gravity of 1,254.

It is employed as local application to relieve rheumatic **or neuralgic pains.**

ETHYLENE-DIAMINE TRICRESOL.

A mixture of 10 parts each of ethylendiamin and tricresol, dissolved in 500 parts of distilled water. It appears as a clear, colorless liquid, becoming slightly yellow on exposure to the air. It has an alkaline reaction, and metallic instruments are not attacked by it when in dilute solution.

ETHYLENE-IMINE. See Piperazine.

ETHYLIDEN CHLORIDE. $CH_3 CH Cl_2$.

Synonyms: Æthylidenum Chloratum ; Chlorethyliden.

Ethyliden chloride is prepared by the action of chlorine on ethyl chloride. It constitutes a colorless ethereal liquid, of an agreeable fruity odor, boiling point being 57^d C. (134.6^0 F.) and specific gravity 1.18.

It is employed as an anæsthetic in minor **operations, producing rapid narcosis of** short duration.

ETHYLIDEN-DI-ETHYL ETHER. See Acetal.

ETHYLIDEN-DI-METHYL-ETHER. See Dimethylacetal.

EUCALYPTEOL.

This antiseptic is obtained by the treatment of eucalyptus oil with hydrochloric acid, producing eucalyptene bichloride, in colorless, scaly crystals, with an odor resembling camphor, and almost tasteless. It is insoluble in water, soluble in ether, chloroform and alcohol, being decomposed by the latter. The dose for adults in capsules is 1 to 1.5 Gm. (15 to 24 grains),

EUCALYPTOL. $C_{10}H_{18}O$.

Synonyms: Cineol, Cajeputol

Eucalyptol is an oxygenated body obtained from the volatile oil of various species of *Eucalyptus*. It is identical with *cajeputol* of oil of cajeput, and *cineol* of oil of worm-seed. Eucalyptol forms a colorless liquid of camphoraceous odor, boiling at 176°C. (348.8° F.); sp. gr. 0.930. It is insoluble in water, but miscible with alcohol, ether and the fatty oils. Eucalyptol, being the active constituent of oil of eucalyptus, is employed in all cases where the latter is applicable, that is as an antiseptic, rubefacient, etc.; it is an effective agent in lung and bronchial affections, being introduced as a spray and inhalation. Its internal dose is 5 drops in capsule or as emulsion.

EUCALYPTO-RESORCIN.

This compound is obtained by warming together molecular quantities of eucalyptol and resorcin; the resulting product is recrystallized from alcohol. It forms a white crystalline powder, which is soluble in alcohol and ether; insoluble in water. Eucalypto-resorcin is an antiseptic, employed for inhalation (in alcoholic solution) in treatment of phthisis.

EUDOXIN.

Synonym: Tetra-iod-phenolphtalein bismuth.

A bismuth salt of nosophen (tetraiodphenolphtalein), containing 52.9% of iodine and 14.5% of bismuth. It is said to be useful in stomachic and intestinal troubles, in doses of 0.2 to 0.5 Gm. (3 to 8 grains) for adults; 0.1 to 0.2 Gm. (1½ to 3 grains) for children of 5 to 10 years old, and for infants in doses up to 0.1 Gm.

EUGENOL. $C_6H_3.(C_3H_5)(O.CH_3)(OH)$.

This is a phenol, which occurs in various volatile oils, particularly those of cloves (80 to 90%), allspice, cinnamon, sassafras and bay. Oil of clove is treated with an excess of liquor sodæ, then shaken with ether to remove the terpenes, the aqueous solution of sodium eugenol is then decomposed by the addition of acid, and the eugenol separates as an oily fluid. It is an aromatic, colorless, oily liquid, which boils at 246°C. (474.8° F.); when exposed to the air it rapidly turns brown. Readily soluble in alcohol, almost insoluble in water; it unites with alkalies, forming soluble salts. Eugenol is a powerful antiseptic, being employed in dental surgery. It has also been recommended in treatment of tuberculosis in doses of 1 to 2 Gm. (15 to 30 grains).

BENZOYL-EUGENOL. Is prepared by the action of benzoyl chloride on eugenol-sodium, occurs in colorless, inodorous crystals, which melt at 70.5° C. (159° F.); insoluble in water, soluble in alcohol, ether and chloroform. This compound is recommended in place of Eugenol.

CINNAMYL-EUGENOL. Is prepared by interaction between cinnamyl-chloride and eugenol sodium; it forms inodorous, colorless crystalline needles, which melt at 90° to 91° C. (194° to 195.8° F.); insoluble in water, soluble in alcohol, ether and chloroform. This compound is likewise recommended for administering in place of Eugenol.

EUGENOL-ACETAMID. Is prepared by the interaction between eugenol-sodium and monochloracetic acid, the resulting eugenol acetic acid is converted into the amide by heating with ammonia. From water it crystallizes in shining plates, from alcohol in fine needles, which melt at 110° C. (230° F.). It is recommended as a local anæsthetic, likewise as an antiseptic in treatment of wounds.

IODO-EUGENOL. Is obtained by the action of iodine on eugenol-sodium. It forms a yellowish-colored, inodorous, insoluble powder, which melts at 150° C. (302°F.). Iodo-Eugenol is employed as an antiseptic.

EUGENOL-BENZOATE. See under Eugenol.

EUGENOL-CINNAMATE. See under Eugenol.

EULYPTOL. See Ulyptol.

EUONYMIN.

A glucoside obtained from the bark of the root of the Wahoo (*Euonymus atropurpureus*). It forms a brownish-colored powder of very bitter taste, slightly soluble in water and soluble in alcohol and ether. Its properties are purgative. Dose 0.03 to 0.2 Gm. (½ to 3 grains).

EUPHORIN. $C_6H_5NH-CO-OC_2H_5$.

Synonym: Phenyl-Urethane.

The esters of carbamic acid $CO \left< \begin{smallmatrix} NH_2 \\ OH \end{smallmatrix} \right.$ are called *Urethanes*, euphorin being a phenyl-ester. It is obtained by the interaction between aniline and mono-chloroformic ethyl-ester, forming a colorless crystalline powder, melting at 49° to 50° C. (120.2° to

122° F.), slightly soluble in cold, more readily in hot water; soluble in alcohol ether and hydroalcoholic mixtures. Euphorin is employed as an antipyretic, antirheumatic and analgesic, in doses of 0.13 to 0.5 Gm. (2 to 8 grains); as an antiseptic it is used in the form of a dusting-powder in the treatment of ulcers, skin diseases, etc.

EUROPHEN. $C_4H_9(OCH_3)C_6H_3 \cdot C_6H_2 \cdot C_4H_9(CH_3)OI$.

Synonym: Isobutyl–Ortho–Cresol–Iodide.

The method of preparation is analogous to that employed in the manufacture of aristol, in which a solution of isobutyl-ortho-cresol in dilute alkali is precipitated by a solution of iodine in potassium iodide. Europhen forms a yellowish, amorphous powder, of aromatic odor, insoluble in water ; easily soluble in alcohol, ether and the fatty oils. It yields iodine to metallic salts. It should be preserved in a dry place, away from the action of light; water and alkalies decompose it. Europhen is employed as an antiseptic, being applied either as a dusting-powder or as a 5 to 10% ointment ; for subcutaneous injection a 3 to 5% solution in olive oil is used.

EXALGINE. $C_6H_5N(CH_3) \cdot (CH_3CO)$.

Synonym: Methyl–Acetanilide.

This compound, a methylated acetanilide, is prepared by the interaction between acetyl-chloride and monomethyl-aniline. It forms acicular needles, which are difficultly soluble in cold water and readily in alcohol and diluted alcohol, melting at 100° C. (212° F.) Exalgine is an antineuralgic, being given in doses of 0.065 to 0.2 Gm. (1 to 3 grains.)

EXTRACTUM LACTIS.

An extract containing the inorganic constituents of milk (in nuclein-like combination), free from milk sugar, casein and albumin, especially adapted as a means of administering calcium. One kilo of this "extract" represents 2000 liters milk.

FAREOL.

A proprietary anodyne and antipyretic.

FELLITIN.

The trade name given by a German manufacturer to a preparation made from ox gall for use in frost bite. Its use is based on the popular application of fresh ox gall to this purpose in Ireland, Russia and certain parts of Germany.

FERCREMOL.

A compound of hæmoglobin and iron, containing 3% of iron. It forms a brown, tasteless powder, which dissolves in weak ammoniated water. The dose is 0.2 to 0.52 Gm. (3 to 8 grains) three times a day.

FERRATIN.

Natural ferratin, a compound of iron found in the liver. May be obtained by extracting the liver of the hog with water; artificially by a patented process. One hundred grammes of egg albumen are placed in a mixture of 21 Cc. of water and 70 Cc. of solution of caustic soda (10%). Twenty grammes of tartrate of iron are dissolved in water, and, if acid, as it generally is, it is neutralized with sodium carbonate. The two solutions are mixed and left for five or six hours, and then transferred to the water bath. The black coloration, due to the formation of sulphide of iron, will disappear towards the end of the process. After cooling, tartaric acid is added to faint acidity. The precipitate formed is dissolved by the addition of a little ammonia, and excess of this is driven off by exposure to the water bath. It is then filtered, and when the filtrate is cold the ferratin is precipitated by a solution of tartaric acid, which should only be added in just sufficient quantity. It is filtered off, washed with water, alcohol and ether, and dried. Ferratin forms a reddish-brown, inodorous and tasteless powder, insoluble in water or dilute acids, but soluble in water, possessing a slight alkaline reaction. The preparation contains about 7% of iron. The commercial article appears in two forms, one being insoluble, as described above, and the other a soluble sodium compound. Ferratin is readily absorbed in the organism without causing the slightest digestive disturbances. Dose, 0.5 Gm. (8 grains); children half this dose.

FERRO-HEMOL. See Hæmol.

FERROPYRIN. $(C_{11}H_{12}N_2O)_3Fe_2Cl_6$.

A compound of three molecules of antipyrin and one molecule of ferric chloride. This compound possesses the combined valuable properties of both its constituents, hence may be employed in the treatment of anæmia, neuralgia, etc. Ferropyrin appears as an impalpable, orange-red colored powder, containing 64% of antipyrin, 12% of iron, and 24% of chlorine. Soluble in 5 parts of water at 15° C., and only 9 parts at 100° C., hence on boiling a cold saturated solution ferropyrin separates, and in the form of ruby-red scales, which melt between 22.° and 225° C. Very soluble in cold methyl-alcohol, from which it separates in orange-red shining scales; also very soluble in alcohol and insoluble in ether. The addition of alkalies or alkali bicarbonates to its aqueous solution causes the precipitation of ferric hydrate.

FERROSINE.

A preparation said to contain iron oxide, 70 to 75%; lime and albumen, 10 to 2 %, and water, etc., 10 to 15%. It occurs either as a granular or fine red powder, which is used as a pigment.

FERRUM ALBUMINATUM. See Iron Albuninate.

FERRUM CASEINATUM.

Synonym: Ferrum Nucleoalbuminatum.

Recommended as a substitute for Ferrum albuminatum, and contains 5.2% ferrous oxide. It is prepared by precipitating a solution of iron lactate with a solution of calcium caseinate. Skimmed milk is diluted with water and the casein precipitated by the addition of acetic acid (avoiding excess). This is collected and repeatedly washed with warm water, finally with alcohol followed by ether. One part of this purified casein is rubbed with one part of calcium carbonate and 100 parts of warm water; the resulting solution of calcium caseinate is filtered and treated with a slight excess of a freshly prepared 1% solution of ferrous lactate. The resulting precipitate of iron casein-ate is at first colorless, but on drying turns a flesh color. It is devoid of odor and taste, and is soluble in water only when made alkaline with sodium carbonate.

FERRUM OXYDATUM SACCHARATUM SOLUBILE. See Iron Saccharated [Oxide.

FERRUM PEPTONATUM. See Iron Peptonate.

FILICIC ACID. (Amorphous). $C_{35}H_{42}O_{13}$.

An amorphous principle obtained from the rhizome of the Male Fern (*Aspidium flix-mas*). This forms a tasteless white powder, which is soluble in alcohol and the fatty oils, melts at 125° C. (257° F.) The anthelmintic properties of male fern extract are ascribed to this principle, which is given in doses of 0.5 to 1 Gm. (8 to 15 grains). If absorbed into the system, amorphous filicic acid is toxic, hence it should not be administered with the fatty oils. The crystalline filicic acid is absolutely inert. (Poulsen).

FLUORESCEIN, or (Resorcin-Phtalein). $C_{20}H_{12}O_5.H_2O$.

Phtalic acid anhydride (75 p.) is fused with resorcin (100 p.), the mass well washed with hot water and crystallized from alcohol. Fluorescein forms a yellowish-red, crystalline powder, insoluble in water, uniting with alkalies to form soluble salts. A 2% alkaline (Na HCO₃) solution is employed in diagnosis of corneal lesions, and detection of minute foreign bodies imbedded in that tissue. Those portions of the cornea which are devoid of their epithelium are colored green, while foreign bodies are surrounded by a green ring. (Straub.)

FLUOROL.

A sodium fluoride which is recommended as an antiseptic of equal value to sublimate, potassium permanganate and formaldehyde. It possesses the advantage of not coagulating albumen. Injections of a 1 to 2 0 solution of fluorol are neither painful nor caustic, and they produce no irritation whatever when applied to the mucous membranes, while they render the latter unfit for the propagation of micro-organisms.

FORMALIN. $H.COH + xH_2O$.

Synonyms: F rmic Aldehyde ; Formol.

This is a concentrated (40%) aqueous solution of formic aldehyde, the latter being obtained by passing the vapors of methyl alcohol over glowing coke or platinum spirals. This solution possesses a pungent odor and neutral reaction, its sp. gr. being from 1.030 to 1.083. Formic aldehyde is a most powerful antiseptic, its solution and vapors being devoid of any deleterious effect upon tissues or objects ; a 1 to 2% solution is sufficiently strong for all purposes. Incompatible with NH₃, bisulphites; also reduces alkali silver and copper salts.

FORMANILIDE. $C_6H_5NH.COH$.

This is obtained by digesting aniline with formic acid, or by rapidly heating it with oxalic acid. It forms colorless prismatic needles, melting at 46° C. (115° F.), readily soluble in water, alcohol, glycerin and the oils. Formanilide is employed as an antipyretic and analgesic in doses of about 0.12 to 0.3 Gm. (2 to 5 grains). When applied to the mucous membrane in powder form, or used in the form of a subcutaneous injection (1 Cc. of a 3% solution), it acts as a local anæsthetic.

FORMIC ALDEHYDE. See Formalin.

FORMOL. See Formalin.

FORMYL CHLORIDE. See Chloroform.

FORMYL-PHENACETINE. See under Phenacetine.

FORMYLPIPERIDIN.

When piperidin is heated to 105° C. in an autoclave, with its own weight of an alkali, there results formylpiperidin and chloroform. The former is an oily liquid which possesses an agreeable, aromatic odor. It boils at 226° to 221° C. (428° to 432° F.), and has its melting point (as a platinum salt) at 171° to 172° C. (339.8° to 341.6° F.) It is soluble in water and in alcohol. On conducting hydrochloric acid **gas** through a solution of formylpiperidin in absolute ether, hygroscopic crystalline **needles** having the structure $C_5H_{10}N.COH.HCl$ are produced.

FORMYL TRIBROMIDE. See Bromoform.

FORMYL TRIIODIDE. See Iodoform.

FOSSILIN.

A name given to a petroleum product similar to petrolatum, vaselin, etc.

FRAXININ.

Synonym: Mannite.

FRUIT-SUGAR. See Diabetin.

GADUOL.

Synonyma: Alcoholic Extract of Cod Liver Oil; Morrhuol.

Brownish-yellow, oily liquid, bitter, acrid taste. Same uses as cod liver oil. **Dose,** 5 to 16 min. in capsules.

GALLABROMOL. See Gallobromol.

GALLACETOPHENONE, $CH_3\text{-}COC_6H_2(OH)_3$.

Synonyms: Alizarin-Yellow C. ; Methyl-keto-trioxybenzene ; Tri-oxy-aceto-phenon.

This derivative of pyrogallol is known commercially under the name of "alizarine-yellow C." It is prepared by interaction between pyrogallol, acetic acid and zinc chloride at 150° C. (302° F.) It forms a pale yellow powder, almost insoluble in cold water, readily soluble in hot water, alcohol, ether and glycerin. Gallacetophenone is employed in dermatology (10% ointment) as a substitute for pyrogallol, which frequently gives rise to toxic symptoms.

GALLANOL. $C_6H_5NH.CO.C_6H_2(OH)_3$.

Synonyms: Gallic Acid Anilide; Gallinol.

This compound, the anilide of gallic acid, is obtained by boiling tannin with aniline. It is a colorless, crystalline solid, with bitter taste, soluble in water, alcohol and ether; possessing marked astringent properties. Gallanol is employed in skin diseases in place of chrysophanic acid and pyrogallol, being less irritating and without poisonous properties. The strength of the ointment varies from 3 to 20%; in some instances it is used as a dusting-powder when mixed with French chalk.

GALLIC ACID ANILIDE. See Gallanol.

GALLICIN. $C_6H_2 <^{(OH)_3}_{COOCH_2}$

A methyl ether of gallic acid, prepared by passing dry hydrochloric acid gas **through** a solution of gallic or tannic acid in methyl alcohol. Occurs either in rhombic **prisms** or fine needles which melt at 206° C., readily soluble in water and alcohol. Recommended as a non toxic, antiseptic dusting powder.

GALLINOL. See Gallanol.

GALLOBROMOL. $C_6Br_2(OH)_3COOH$.

Synonyms: Di-bromo-gallic acid ; Gallabromol.

This compound, obtained by the action of bromine on gallic acid, occurs in fine white needles, almost insoluble in cold, but readily soluble in hot water, also in alcohol and ether. It is used in neurasthenia and similar complaints as a sedative in place of the alkaline bromides, the dose being 1 to 10 Gm. (15 to 150 grains.) Also employed in cystitis and epididymitis (by irrigation with 2 to 4% solutions), and in eczema madidum and crustosum (1 to 2% solutions, powders or ointments).

> ℞ Gallobromol........................... 3 Gm.
>   Dist. water........................... 200 "

For injection 4 to 5 times a day (in gonorrhœa, **cystitis** and epididymitis, **and as a** lotion in eczema)

GELATOL.

An ointment **base composed** of a mixture of oil, glycerin, gelatin and water.

40   THE NEWER REMEDIES.

GLACIALIN.

A mixture of borax, boric acid and sugar.

GLONOIN. Nitroglycerin (see U. S. P.).

GLUCIN.

The sodium salt of amido-triacin-sulphonic acid, obtained by the action of alde-
hydes upon chrysoidine and conversion of the condensation products into their mono
and disulphonic acids. As regards sweetening power, glucin is inferior to saccharin,
being about 100 times sweeter than cane sugar.

GLUCUSIMIDE. See Saccharin.

GLUSIDE. See Saccharin.

GLYCERIN—PHOSPHORIC ACID.

This compound, a glycerin ester of phosphoric acid, is prepared by interaction of
phosphoric acid and glycerin in the presence of dehydrating agents. It is a yellowish
inodorous, oily liquid, of acid taste, soluble in water and alcohol; its calcium salt being chiefly employed. Used subcutaneously in doses of about 0.25 Gm. (3.8 grains) to
increase the amount of phosphorus in the organism of neurasthenics and those afflicted with nervous disorders. The potassium, sodium and calcium salts are employed
in convalescence from influenza and other infectious diseases, as also in nervous
asthenia of various origin. The sodium salt is administered hypodermically in daily
doses of 0.20 to 0.25 Gm. (3 to 4 grains.)

GLYCOLINE.

A purified petroleum oil, for use in atomizers.

GLYCOZONE.

A thick, syrupy liquid, which is made by saturating glycerin with ozone. It is administered in teaspoonful doses, diluted with water, in treatment of dyspepsia, etc.

GLYMOL.

A proprietary preparation, claimed to be a liquid hydrocarbon, of neutral reaction,
obtained from crude petroleum; specific gravity, .885, at 60°F. Employed in nasal and
bronchial diseases, gynæcological practice, etc.

GOLD MONOBROMIDE.

Yellowish-gray, very friable mass, insoluble in water. Antiepileptic, antisyphilitic
Dose, 1-8 to 1-5 grain. Anodyne (migraine, etc.), 1-20 grain twice daily before meals.

GOLD TRIBROMIDE.

Soluble in water. Therapeutics and dose same as mono-bromide.

GUAIACOL. $C_6H_4(OCH_3)$ (O II) (1:2).

Synonyms: Methyl-Pyro-Catechol: Methyl-Pyrocatechin.

Beechwood tar creosote, which consists of a mixture of guaiacol, cresols and creosol,
is fractionated, collecting that portion which comes over between 200° and 205° C.; this
product on treatment with alcoholic potassium-hydrate, yields potassium-guaiacol,
which, when decomposed with dilute acids, liberates guaiacol. When pure, guaiacol
forms a crystalline solid, which melts at 28.5°C. (83.3°F.), and boils at 205.1°C. (401.1°F.);
purified liquid creosote is a colorless refractive liquid, of agreeable aromatic odor,
having a specific gravity of about 1.143; soluble in 85 parts of water, readily in all
proportions in alcohol and ether. Commercial guaiacol (generally of synthetic origin),
does not contain more than 91% of pure guaiacol, its gravity is lower and color darkens
on exposure to air. Guaiacol is a valuable remedy in phthisis, being given in doses of 1
to 2 minims, which may be increased to 20 minims or more. It readily combines with
acid radicals forming crystalline compounds, among which are :

CREOSOTE CARBONATE (Creosotol). This compound is analogous to guaiacol
carbonate, but is prepared directly from beech-wood creosote, instead of guaiacol. It
forms a thick, brownish, inodorous oil, insoluble in water. Creosotol is preferred to
creosote for internal administration since it is readily absorbed and free from all disturbing symptoms which accompany creosote itself. Dose is 3 to 15 minims.

GUAIACOL BENZOATE. See Benzosol.

GUAIACOL BINIODIDE. Is prepared by precipitating an aqueous solution of **sodium guaiacol** with a solution of iodine in potassium iodide. It forms a reddish-**brown powder,** possessing an odor of iodine, soluble in alcohol and the fatty oils. **Nothing definite** is known as to its dose.

GUAIACOL CARBONATE. [CO₃(C₆H₄OCH₃)₂]. The di-guaiacol ester of carbonic acid, is formed by the action of phosgene gas on guaiacol sodium. This forms an inodorous, neutral, crystalline powder, (containing 91.5% of guaiacol), insoluble in water, slightly so in alcohol, glycerin and the oils; melts at 65° C. (149° F.). The irritation produced by guaiacol, as well as creosote, has added to **the popularity of this** salt, which does not disturb the digestive functions, for, being **insoluble,** it passes unchanged through the stomach into the intestines, where it is **split up.** **The dose is** 0.3 to 0.5 Gm. (5 to 8 grains), gradually increasing to 5 Gm. (75 **grains) daily.**

GUAIACOL CARBONIC ACID. (C₆H₃(OH)(OCH₃) COOH+2H₂O. Is prepared by passing carbonic-acid **over** sodium-guaiacol heated to 100° C. (212° F), the resulting product on treatment with acids yields the free acid. This forms a white, crystalline, inodorous powder of bitter taste; slightly soluble in water, readily in alcohol and ether, melting at 150° C. (302° F.). Guaiacol carbonic acid and its alkali salts have been recommended as antiseptics and antirheumatics. This compound should not be confused with Guaiacol Carbonate.

GUAIACOL CINNAMATE or *STYRACOL.* C₆H₅.CH=CH.CO₂C₆H₄.OCH₃. Is the cinnamic ester of guaiacol. It is prepared by warming a mixture of guaiacol and cinnamyl chloride in molecular proportions. This compound forms colorless needles which melt at 130° C. (266° F.), insoluble in water, readily soluble in alcohol. Styracol is employed in catarrhal affections of the digestive organs, also in the treatment of phthisis.

GUAIACOL SALICYLATE or *GUAIACOL-SALOL.* C₆H₄(OH) COO—C₆H₄-(OCH₃). A compound analogous to salol, is prepared by the action of phosphorus oxychloride on a mixture of guaiacol sodium and sodium salicylate. It forms a white, inodorous, tasteless, crystalline powder, melting at 65° C. (149° F.), almost insoluble in water, soluble in alcohol and ether. It is administered to phthisical patients to aid digestion, also as an intestinal antiseptic in doses of 1 Gm. (15 grains).

GUAIACOL PHOSPHATE. PO (C₆H₄ O CH₃O)₃. This salt is prepared by H. Dubois by making **a** solution of guaiacol in soda lye, cooling, and then adding phosphorus oxychloride, drop by drop, in somewhat more than the theoretical quantity. After standing five or six hours an oily layer of the phosphate collects on the bottom of the vessel, which soon crystallizes, and is then purified by repeated washing with alcohol. It crystallizes in hard colorless tables, melting at 98°C. It is insoluble in water, alcohol and petroleum ether, and easily soluble in chloroform and acetone.

GUAIACOL SUCCINATE. This new ester of guaiacol may be prepared either by treating a mixture of guaiacol and succinic acid with a definite quantity of phosphorus oxychloride, or, preferably, by treating an aqueous soda solution of guaiacol, cooling it the while with succinyl chloride. It has the formula C₄H₄O₄(C₆H₄OCH₃)₂. It crystallizes in fine needles with a silken luster, melting at 136° C. It is insoluble in water, slightly soluble in ether and alcohol, and readily so in chloroform.

OLEO-CREOSOTE, the oleic ester of creosote, prepared by combining creosote **with oleic** acid by means of phosphorus trichloride. This is a yellow, oily liquid (35% **creosote),** insoluble in water and nearly so in alcohol, soluble in ether and in oils. It is used **as an** antiphthisic in doses of 15 to 16 minims.

GUARANIN. See Caffeine.

GYMNEMIC ACID. C₃H₅₅O₁₂.

The active principle prepared from the leaves of *Gymnema silvestre.* It forms a greenish white powder, of an acid astringent taste, sparingly soluble in water, easily in alcohol. It produces a temporary ageusia to sweet and bitter tastes. Before partaking of bitter medicines, the mouth is rinsed out with a 1% per cent hydroalcoholic solution.

GYNOCARDIC ACID.

The active principle from the oil of the seeds of *Gynocardia odorata.* It forms a yellowish, unctuous solid, melting at about 30° C. (86° F.); it has a burning and acrid taste, and marked odor. Used internally and externally in treatment of leprosy and syphilis, and of gouty and rheumatic affections. Dose, ½ to 8 grains; externally, as liniment with oil (1: 10 to 20).

GYNOCYANAURIDZARIN.

A principle (crystalline) obtained from the *Gynocardia lancifoliata.* It forms unstable crystals, insoluble in the usual solvents, soluble 1 in 5000 in olive oil, the solubility being increased to 1 in 80 by the addition of cinnamic alcohol. This solution, which contains 0.00001 Gm. (1-6500 grain) to the minim, is recommended in doses of 8 to 20 minims in various forms of tubercular affections.

HÆMALBUMIN.

A predigested iron albuminate, one gramme containing all the constituents which are found in 6 grammes of fresh healthy blood, with exception of fibrin and such products as urea, kreatinin, etc. Also 1 Gm. of Hæmalbumin, aside from the readiness with which it is absorbed, is equal to 25 Gm. of Liquor Ferri Albuminati. For infants, one gramme dissolved in hot water, with sufficient sugar, is used. For adults the same quantity may be given in dry powder form several times daily.

HÆMATOGEN.

A yellowish powder, containing 7% of iron, or a liquid obtained by adding ferric citrate and acetic acid to an alkaline solution of albumen. This is employed in treatment of rachitis and scrofulous conditions, also as a tonic for anæmics. Dose of the liquid is 1 to 4 teaspoonfuls, according to age.

HÆMOGALLOL.

A ferruginous blood preparation, obtained by oxidizing the hæmoglobin of the blood (defibrinated blood) by the action of pyrogallol, thus furnishing a compound which is easily assimilated and supplies those constituents of the blood which are found lacking in chlorosis and anæmia. It forms a red-brown powder, insoluble and tasteless, being given in doses of 0.2 to 0.5 Gm. (3 to 8 grains).

HÆMOGLOBIN.

The red coloring matter of the solid principles of the blood. It forms a red powder soluble in water and used in treatment of anæmia and chlorosis in daily doses of 5 to 10 Gm. (75 to 150 grains), taken usually in wine.

HÆMOL (FERRO).

A preparation closely allied to hæmogallol, being obtained in the same manner, except that zinc-dust is employed as reducing agent. This a dark brown insoluble powder, administered as a tonic in doses of 0.1 to 0.5 Gm. (1½ to 8 grains).

BROMHÆMOL or *BROMATED HÆMOL*, contains 2.7 per cent of bromine. Recommended in epilepsy.

CUPROHÆMOL or *COPPER HÆMOL*, a dark brown powder, containing 2 per cent. of copper in non-irrigating form, and therefore an eligible succadaneum for the older copper compounds in tuberculosis, scrophulosis, etc. Dose is 0.10 to 0.15 Gm. (1½ to 2 grains), three times daily in pill form.

IODOHÆMOL contains 16.6 per cent. of iodine.

MERCUROIODOHÆMOL is a brownish-red powder, containing 12.35 per cent of metallic mercury and 28.68 per cent. of iodine. Antisyphilitic in doses of 0.08 Gm. (1¼ grains), 6 to 10 times daily.

ZINCOHÆMOL is a brown, almost insoluble powder, containing 1 per cent. of zinc. Recommended as a mild antidiarrhoeic in doses of 0.5 Gm. (8 grains), three times daily.

HELCOSOL. See Bismuth Pyrogallate.

HELENIN. C_6H_8O. See Alantol.

A stearoptene obtained from the root of *Inula Helenium* (Elecampane Root) by exhaustion with alcohol and precipitating the resulting extract by pouring into water. It forms white, acicular crystals, which melt at 110° C. (230° F.), insoluble in water, readily soluble in hot alcohol, also in ether and the oils. Helenin is employed in treatment of whooping cough, bronchitis and tubercular coughs in doses of 0.01 Gm. (1·6 grain)

HELIOTROPIN. See Piperonal.

HEMATIN-ALBUMIN.

A preparation of blood, consisting principally of dried albumins, holding a large amount of iron. It is a dark-brown powder, odorless and almost tasteless. One pound of hematin-albumin is said to contain the albumins of about 6 pounds of blood. Dose for adults is 1 to 2 teaspoonfuls three times daily.

HEXA-METHYLENE TETRAMIN. See Urotropin.

HOMATROPINE. $C_{16}H_{21}NO_3$.

Synonym : Oxy-toluol-tropine.

An artificial alkaloid obtained from tropine mandelate, prepared synthetically by Ladenburg from tropic acid and tropin, the two derivatives of atropine. This forms colorless, very hygroscopic crystals, slightly soluble in water. Its action is like that of atropine, but less persistent and weaker, causing, when applied to the eye, rapid dilation of the pupil, which passes off sooner than that of atropine. Also given internally in treatment of the night-sweats of phthisis. The maximal internal dose is 0 001 Gm. (1-64 grain); as application, in 1 per cent. solution. The salts of the alkaloid are preferred.

HOMATROPINE HYDROBROMATE, ($C_{16}H_{21}NO_3$. H Br), because of its ready solubility and non-hygroscopic nature, is preferred to the alkaloid. The other salts of homatropine are the *hydrochlorate, salicylate* and *sulphate*.

HYDRACETINE. $C_6H_5NH-NH-CH_3CO$.

Synonyms: Pyrodine ; Acetyl-phenyl-hydrazine.

This compound may be looked upon as hydrazine H_2N-NH_2, in which a hydrogen in each of the NH_2 groups is replaced by a monovalent radical, one being a phenyl (C_6H_5), the other being an acetyl (CH_3CO) group. or it may be considered as being the acetyl derivative of phenyl-hydrazine. It is obtained by heating together acetic anhydride and phenyl hydrazine. Hydracetine occurs in colorless, inodorous and tasteless crystals, which melt at 128.5° C. (263.5° F.), soluble in 50 parts of water and readily so in alcohol. Its properties are those of an antipyretic and antirheumatic, in doses of 0.05 to 0.1 Gm. (3-5 to 1½ grains). Care should be taken in administering this remedy, because of its powerful and toxic properties.

HYDRASTINE. $C_{21}H_{21}NO_6$.

An alkaloid obtained from the rhizome of *Hydrastis canadensis*. It occurs in yellowish white crystals which melt at 132° C. (269.6° F.), of intensely bitter taste, insoluble in water, readily soluble in alcohol and ether. Hydrastine is employed in metrorrhagia, also as a tonic and antiperiodic in doses of 0.015 to 0.03 Gm. (¼ to ½ grain). It is not used externally, because of its insolubility.

HYDRASTINE HYDROCHLORATE forms a pale yellow, crystalline powder, of very bitter taste, readily soluble in water and alcohol. It is employed in gonorrhœa, conjunctivitis, leucorrhœa, etc., externally in various dermal affections in a one per cent ointment or lotion.

HYDRASTININE. $C_{11}H_{11}NO_2$.

This is obtained as the oxidation product of hydrastine by nitric acid. It forms acicular crystals, melting at 116° to 117° C. (240.8° to 242.6° F.), insoluble in water, readily soluble in alcohol and ether.

HYDRASTININE HYDROCHLORIDE is usually employed in medicine because of its ready solubility. It occurs in yellow crystals which melt at 205° C. (401° F.). It is employed as a uterine hæmostatic, also in dysmenorrhœa, metrorrhagia, etc., in doses of 0.025 Gm. (⅜ grain). As a subcutaneous injection ½ to 1 Cc. of a 10% aqueous solution once daily.

HYDRAZINE. See Diamine.

HYDROCHINON. See Hydroquinone.

HYDROCINNAMIC ACID. See Beta-phenyl-propionic acid.

HYDRONAPHTHOL.

An antiseptic and disinfectant, said to be obtained from beta-naphthol by reduction, in which a hydrogen atom is replaced by the hydroxyl group. Usually given in keratin or salol-coated pills containing 0.1 to 0.2 Gm. (1½ to 3 grains). For external use, in a one per cent. solution.

HYDROQUINONE. $C_6H_4(OH)_2$ (1:4).

Synonyms: Hydrochinon ; Para-dioxybenzol ; Para-diphenol.

This body is an isomer of resorcin, being prepared by the oxidation of aniline with chromic acid mixture. It forms colorless, hexagonal prisms, which melt at 169° C, (336.2° F.), difficultly soluble in cold water, readily so in hot water, in alcohol and in ether. Hydroquinone is used as an antiferment, antiseptic and antipyretic ; as antipyretic its dose is 1 Gm. (15 grains); as an injection or wash in 10% solution.

Hydroquinone is largely employed as a photographic developer.

HYDROXYLAMINE HYDROCHLORIDE. $NH_2OH.HCl$.

Synonym: Oxy-ammonium chloride.

Hydroxylamine may be regarded as ammonia, NH_3, in which a hydrogen atom is replaced by the hydroxyl group OH. This base is obtained by interaction between sulphurous and nitrous acids at low temperature. Hydroxylamine hydrochlorate forms colorless, hygroscopic, crystalline plates, readily soluble in water, glycerin and alcohol. It is characterized by its great reducing power, precipitating such metals as gold, silver and mercury from their solutions; it likewise reduces Fehling's solution. This compound is employed as an antiseptic in place of chrysarobin, pyrogallol and anthrarobin in treatment of skin diseases as a 1-10 to 4-10 per cent solution.

HYOSCINE. (Scopolamine, *Schmidt*). $C_{17}H_{21}NO_4$.

This amorphous alkaloid occurs, along with atropine and hyoscyamine, in the various solanaceous plants, particularly the seeds of *Hyoscyamus niger*. Hyoscine is identical with, or, according to Schmidt and Hesse, it is *Scopolamine* ($C_{17}H_{21}NO_4$), as obtained from the roots of the *Scopolia atropoides ;* commercial hyoscine being scopolamine. Among the various salts employed are the hydrobromate, hydrochlorate, hydro iodate and sulphate. It is stated that a solution of scopolamine, 1 to 1000, is five times stronger than the same solution of atropin.

HYOSCINE HYDROBROMATE ($C_{17}H_{21}NO_4.HBr+3H_2O$) occurs in colorless, permanent, odorless, acrid crystals. It is employed as a hypnotic and sedative in various mental diseases, also as an antaphrodisiac, antisialagogue and mydriatic. Its dose as a hypnotic in insanity is 0.02 Gm. (1-30 grain), as sedative 0.0004 to 0.0006 Gm. (1-150 to 1-100 grain). Subcutaneously, as hypnotic, 0.0004 to 0.0006 Gm. (1-150 to 1-100 grain), as sedative 0.0002 to 0.0003 Gm. (1-300 to 1-200 grain). As a mydriatic a 1% solution is used. Antidotes the same as for atropine.

HYOSCINE HYDROIODATE. See Iodic Acid.

HYOSCYAMINE. $C_{17}H_{23}NO_3$.

An alkaloid which occurs with hyoscine and atropine in the seeds and leaves of the *Hyoscyamus Niger*, also found in roots of the *Scopolia Atropoides* and *Japonica*, also in the leaves of the *Duboisia Myoporides*, etc. It forms white, silky, permanent crystals, melting at 108.5° C. (227.3° F.), almost insoluble in water, readily soluble in alcohol and ether. The action of hyoscyamine is like that of atropine, but it is chiefly employed as a hypnotic in mental disorders, as an anodyne and antispasmodic in asthma, epilepsy, colics, etc. Its usual dose is ½ to 2 Mg. (1-120 to 1-30 grain); as hypnotic for the insane 0.0075 to 0.015 Gm. (⅛ to ¼ grain.

Among the various soluble salts employed are the *hydrobromate*, *hydrochlorate* and *sulphate*.

HYPNAL. (Monochlorantipyrine.) See under Antipyrine.

HYPNOACETIN. $CH_3CO-NH-C_6H_4-OCH_2-COC_6H_5$

Synonym: Acetophenon-acetyl-para-amido-phenol-ether.

This forms pearly scales which melt at 16.0° C. (320° F.), soluble in alcohol and acetic ether. Recommended as a hypnotic in doses of 0.2 to 0.25 Gm. (3 to 4 grains.)

HYPNONE. $C_6H_5-CO-CH_3$.

Synonyms: Acetophenone; Methyl-phenyl-ketone

This is a mixed ketone obtained by the dry distillation of a mixture of calcium acetate and benzoate. Hypnone is a colorless, oily fluid, of peculiar odor and pungent taste. Its sp. gr. is 1.032, and when exposed to the temperature of 14° C. (57.2° F.), it solidifies. Only slightly soluble in water, but readily miscible with alcohol, ether and the fatty oils. It is employed as a hypnotic in doses of 0.05 to 0.2 Gm. (7-10 to 3 grains), or 1 to 3 minims.

IATROL.

Synonym: Oxy-iodo-methyl-anilide.

Nothing is known concerning the preparation of this compound, which is described as being an inodorous, non-poisonous antiseptic, designed to replace iodoform.

ICHTHYOL. $C_{28}H_{36}S_3O_6$ (N H_4)$_2$.

Synonym: Ammonium-ichthyol-sulphonate.

A bituminous mineral of Tyrol, which is rich in fossilized remains of aquatic animals, is subjected to dry distillation, yielding a dark, oily distillate; this is treated with an excess of sulphuric acid, by which ichthyol-sulphonic acid is formed; this product on being purified and neutralized with ammonia yields ammonium ichthyol-sulphonate. Ichthyol forms a thick brownish liquid, of bituminous odor and taste, containing 15% of easily assimilable sulphur; its sp. gr. is 1.006; soluble in water, glycerine, a mixture of equal parts of alcohol and ether, and the oils. It is employed externally in various skin diseases, rheumatism, inflammatory diseases, and in gynaecological practice; internally it is given for various affections of the digestive and intestinal tract. also in treatment of scrofula, syphilis, etc. As external application, from 5 to 50% ointment or solution is used; in gonorrhœa 1 to 3% solutions are employed; the internal dose is 0.2 to 0.6 Gm. (3 to 10 m.). 3 times daily in pills or capsules

Among the various other salts of ichthyol sulphonic acids, are the ichthyol-sulphonates of sodium. magnesium, zinc and mercury. These are black, tarry-like masses, the magnesium salt making the best pill, while the zinc salt is best for injections. Ichthyol is incompatible with strong alcoholic liquids and acids.

INDOPHENINE. See under Phenacetine.

INGESTOL.

An opalescent, yellow liquid, recommended for the treatment of acute and chronic disturbances of the stomach and intestines, also sea-sickness. As far as known, it contains magnesium, potassium and sodium sulphates, sodium chloride, ferric chloride, alcohol and water.

IODANTIFEBRIN. C_6H_4 INII C_2H_3O.

Synonym: Iodacetanilid.

Prepared by the action of iodine on acetanilid. **It forms a crystalline powder,** insoluble in water. As far as its action is **concerned it is almost inert.** Nothing is known concerning its properties.

IODIC ACID AND ITS COMPOUNDS.

IODIC ACID, HIO_3, is extensively employed for the reduction of chronic glandular enlargements and goitre. ½ dr. of a 2% solution being injected into the affected parts.

ALKALI IODATES
SODIUM **IODATE, NaIO$_3$**
POTASSIUM IODATE, KIO$_3$
Owing to the ready liberation of iodine when brought into contact with mucus surfaces, they are employed, in dilute solution, 2 to 5% in various affections of these tissues.

SILVER, ZINC AND STRONTIUM IODATES. $AgIO_3, Zn(IO_3)_2, Sr(IO_3)_2$ are insoluble. Employed externally in various affections of mucus surfaces. The silver salt is also given internally as an intestinal astringent in doses of 1-0 to ¼ Gr. .005 to .01 Gm. best administered in pill form.

LITHIUM IODATE. $LiIO_3$ is employed in doses of 1½ gr. (0.1 Gm.), subcutaneously injected, in cases of renal colic, or in cases of uric acid diathesis, or in chronic gout. In the latter case it is best employed in pill form in doses of 1½ to 3 Grs. (0.1 to 0.2 Gm.)

MERCURIC IODATE. $Hg(IO_3)_2$ in form of a double salt with KI, is successfully employed during all stages of syphilis; no salivation or other side effects becoming apparent. It is best used in subcutaneous injections in doses of 1-6 gr. (.01 Gm.).

QUININE HYDRO-IODATE. $C_{20}H_{24}N_2O_2HIO_3$ is recommended as an excellent nervous sedative and antineuralgic. It may be given internally **or** hypodermically in doses of 1 to 1½ grs. (0.06 to 0.1 Gm.).

STRYCHNINE HYDRO-IODATE. $C_{21}H_{22}N_2O_2HIO_3$ may be used **in all cases** where strychnine is indicated, in doses as high as 1-12 gr. (.005Gm.).

CODEINE HYDRO-IODATE. $C_{18}H_{21}NO_3HIO_3$ is employed **for the same purpose and in the same doses as other salts** of codeine, its action being **somewhat more energetic.**

HYOSCINE HYDRO-IODATE. $C_{17}H_{21}NO_4HIO_3$ may be **advantageously substituted** in place of the chloride, iodide, or bromide; it may be **employed subcutaneously** or internally, in either case its action is more intense and its **dose should be smaller** than that of the other salts.

ATROPINE HYDRO-IODATE. $C_{17}H_{23}NO_3HIO_3$ is employed **in ophthalmic practice in ½ to 1½% solution,** with good results.

IODINE TRI-CHLORIDE. ICl$_3$.

Prepared by passing dry chlorine gas over dry iodine which is warmed; the iodine trichloride which forms sublimes in the cooler portions of the apparatus. Orange to yellow, hygroscopic needles, which melt at 25° C. (77° F.), fuming on exposure to the air; when warmed it decomposes into iodine monochloride and chlorine. Soluble in alcohol and water; when dissolved in a large excess of the latter decomposition ensues. Iodine terchloride is a powerful antiseptic and disinfectant (1:1000), its value depending upon the liberation of chlorine, which is rendered still more active by the presence of iodine. When combined with aqua ammonia, iodide of nitrogen is formed. With alkalies iodine is precipitated; when combined with organic substances, iodine is liberated.

IODO-CAFFEINE. See under Caffeine.

IODO-CASEIN.

A new antiseptic and iodoform substitute. **It is a yellowish powder, with a faint** iodine odor.

IODO-HEMOL. See Hæmol.

IOD-IODOFORMIN.

A light-brown insoluble powder, melting at 200° C. (392° F.) Recommended as an iodoform substitute.

IODOFORMAL.

A yellow powder having a strong odor of cumarin, possessing an advantage over iodoform in its extreme lightness and absence of odor. It is insoluble in water and ether, melts at 126° C. (262.4° F.), and yields iodoform when treated with acids. It is distinguished from iodoform in that it yields iodine by the action of sulphuric acid. Iodoformal is intended to be employed in place of iodoform.

IODOFORMIN.

Synonym: Hexa-methylene-tetramin-iodoform.

This compound, containing 75% of iodoform, may be prepared by rubbing together in a mortar 26 Gm. of hexamethylenetetramin and 74 Gm. of iodoform with a little absolute alcohol until a dry powder results. This is a harmless, inodorous compound of iodoform, intended to be used in all cases where iodoform is indicated. It melts at 128° C. (262.5° F.) Iodoformin, when applied to a moist surface, breaks up into its constituents; that is, iodoform is liberated. Maximum dose is 0.25 Gm. (4 grains.)

IODOFORMIN-MERCURY.

A colorless to pale yellow insoluble powder, recommended as an antiseptic.

IODOGEN.

A mixture of charcoal and potassium iodate (KIO_3) formed into cones; these, when ignited, liberate iodine, which serves as an aerial disinfectant.

IODO-PHENIN (Iodo-phenacetin) $C_{20}H_{25}I_2N_2O_4$.

An iodine substitutions product of phenacetine obtained by the action of iodine in potassium iodide on a solution of phenacetin in hydrochloric acid. Iodophenin forms a brownish powder or crystals, containing 25% of iodine, having an iodine-like odor, soluble in alcohol and glacial acetic acid, decomposed by water. It is used externally like iodine.

IODO-PHENO-CHLORAL.

This is a mixture of equal parts of tincture of iodine, carbolic acid, and chloral hydrate, forming a brown-colored fluid, which is recommended as a parasiticide in certain skin diseases.

IODOPYRINE. See under Antipyrine.

IODO-THEINE.

A combination of hydriodic acid and theine (caffeine) forming a crystalline or amorphous white powder, which is decomposed into its constituents by water. Used to increase the systolic action and arterial pressure of the heart. Dose, 0.13 to 0.5 Gm. (2 to 8 grains).

IODOL. C_4I_4NH.

Synonym: Tetraiodopyrrol.

To a solution of pyrrol (1 part) in alcohol (10 parts), a solution of iodine (12 parts) in alcohol (240 parts) is added, and allowed to stand 24 hours; on mixing this product with four times its volume of water, iodol separates in yellow flocks. Iodol, which contains 89% of iodine, forms a pale yellow, inodorous, tasteless powder, insoluble in water, soluble in 3 parts of alcohol, 15 parts of ether, 50 parts of chloroform and 15 parts of oil. Iodol was introduced as a substitute for iodoform, possessing the advantage of being inodorous and non-toxic.

IODOL-CAFFEINE ($C_8H_{10}N_4O_2+C_4I_4NH$) is a crystalline compound, made by the interaction between molecular weights of iodol and caffeine in concentrated alcoholic solution. It forms an inodorous, tasteless, crystalline powder, insoluble in the usual solvents. Used as an antiseptic like iodol, of which it contains 74.6 per cent.

IODO-THEOBROMINE. See under Caffeine.

IRON ALBUMINATE.

Synonym: Ferrum Albuminatum.

This is a compound of ferric chloride and albumen, forming golden-yellow, transparent scales, which are soluble in water. See Dispensatories.

IRON CASEINATE. See Ferrum Caseinatum.

IRON NUCLEOALBUMINATE. See Ferrum Caseinatum.

IRON OXIDE SACCHARATED, SOLUBLE.

Synonym: Ferrum Oxydatum Saccharatum Solubile.

This is a red-brown colored powder, of sweet taste, soluble in 20 parts of hot water. For method of preparation, see Dispensatories.

IRON PEPTONATE.

Synonym: **Ferrum Peptonatum.**

This forms red-brown scales, which are soluble in water. This is a compound of ferric chloride and peptone (digested albumen). See Dispensatories.

β-ISOAMYLENE. See Pental.

IZAL. See under Cresol.

JEQUIRITIN. See Abrin

JESSANODINE.

A proprietary antiseptic and analgesic.

KAIRIN. See under Chinolin.

KAIROLIN. See under Chinolin.

KAPUTINE.

This is said to be merely a colored Acetanilid.

KEFIR (MATZOON; KOUMYS).

These liquors are prepared by the action of various ferments on milk, in some cases, mare's milk, and in other cases, cows milk being used. These preparations possess an undoubted value in all debilitating diseases, and in cases of obstinate vomiting are often well borne. They are very readily assimilated and rapidly increase the body weight after disease. Kefir, owing to its nutritive value, its agreeable taste, its ready assimilation, and its property of assisting digestion in general, has been suggested as a vehicle for the administration of various drugs. It is claimed that the drugs are not simply held in suspension but are held in solution in a stable form. The assimilation of these drugs is no doubt aided by administering them in such combination.

KREOSOTAL KEFIR (Kreosot-Carbonate-Kefir) is found on the market, containing in—

No. 1.— 1 Gm. of kreosotal to the bottle.
2.— 3 " " " " " "
3.— 5 " " " " " "
4.—10 " " " " " "

GUAIAKOL CARBONATEKEFIR also occurs in four combinations, viz:

No. 1.—0.5 Gm. of guaiakol carbonate to the container.
2.—1.0 " " " " " "
3.—1.5 " " " " " "
4.—2 " " " " " "

The above-mentioned preparations of kefir are very useful in the various pulmonary complaints, scrofulous disorders, chronic gastritis, and in hepatic and renal diseases in general. They are best administered at the rate of one bottle per diem, beginning with the lowest number.

ARSENICAL KEFIR. This compound consists of a combination of kefir with Fowler's solution, and contains in—

No. 1.—3 minims of Fowler's sol. to each container.
2.—4 " " " " "
3.—5 " " " " " "
4.—6 " " " " " "

In neurasthenic conditions, chorea, hysteria, various skin diseases, and in all conditions where the arsenical treatment is indicated, this compound is very valuable, far surpassing all other preparations of arsenic.

IODKEFIR consists of sodium iodide combined with kefir, as follows:

No. 1.—0.5 Gm. of sodium iodide, per container.
2.—1.0 " " " " " "
3.—1.5 " " " " " "
4.—2.0 " " " " " "

The permanence of this combination is doubtful. It is recommended in all cases where preparations of iodine are indicated.

KEPHALINE.

A proprietary headache remedy.

KLINOL.

A proprietary antipyretic and analgesic.

48 THE NEWER REMEDIES.

KOSIN. $C_{31}H_{38}O_{10}$.

Synonyms: Koussein ; Kussin ; Kosein.

A bitter principle, isolated from the flowers of *Hagenia abyssinica*, Willd. This forms inodorous, tasteless, yellow-colored crystalline needles, which melt at 142° C (287.6° F.), insoluble in water, readily soluble in alcohol, ether and the alkalies. Kosin is employed as an anthelmintic and tæniafuge in doses of 1 to 2 Gm. (15 to 30 grains).

KOSOTOXIN.

This body is an active principle of Koso flowers. It is a yellowish amorphous powder, according to Leichsewring, melting at 80 degrees, to which the provisional formula $C_{26}H_{34}O_{10}$ has been assigned. It is soluble in alcohol, ether and chloroform, and has a considerable physiological action. It is a strong muscle poison, but has very little action on the central nervous system.

KOUMYSS. See Kefir.

KOUSSEIN. See Kosin.

KRESAPOL. See under Cresol.

KRESIN.

A mixture of a solution of cresylic acid and a solution of sodium-oxy-acetate. It forms a brownish fluid miscible with water and alcohol. A 1 per cent solution is recommended as a general disinfectant.

KRYSTALLOSE.

Name given by Fahlberg, List & Co. to their new sodium-saccharin preparation, or water-soluble crystalline saccharin. It is said to be absolutely free from all contaminations and to be more than 500 times sweeter than sugar.

KUSSIN. See Kosin.

LABORDINE.

A secret remedy, forming a greyish powder, slightly soluble in water, soluble in alcohol. It is stated to contain acetanilid, caffeine, saccharin and possibly a little apiol.

LACTOCINE.

The active principle of the concrete juice of the *Lactuca virosa*. It forms white scales, soluble in 60% of water. Used as a sedative and hypnotic in doses of 1 to 5 grains.

LACTOL, OR LACTO-NAPHTHOL.

A lactic ester of b.-naphthol, similar to benzo naphthol. In the organism it is split up into lactic acid and b.-naphthol, hence is used as an intestinal antiseptic. It forms a colorless and tasteless powder. Dose, about 0.25 to 0.5 Gm. (3.5 to 8 grains).

LACTOPHENINE. See under Phenacetine.

LACTYL-PHENETIDINE. See under Phenacetine.

LACTYLTROPEINE $(C_8H_{14}NO\text{-}CO.CH(OH)\text{-}CH_3)$.

A compound obtained by the action of lactic acid on tropeine. Occurs as white crystalline needles, readily soluble in water, alcohol, ether, etc., melting at 75° C. (167° F.). It appears to have a tonic action on the heart and respiration.

LÆVULOSE. See Diabetin.

LAIFAN.

A Chinese remedy against neuralgia. It is a crude borneol containing water, probably identical with Ngai camphor described by Flückiger, and obtained from *Blumea balsamifera*. It comes into commerce in the shape of a thick paste, showing numerous crystals, put up in earthen pots containing about 6 ounces.

LAMIN.

An alkaloid obtained from the flowers of *Lamium album*. Employed in the form of a sulphate or hydrochloride in subcutaneous injections as a powerful hæmostatic.

LANICHOL.

A specially prepared and purified fat of the wool of sheep. Does not differ essentially from Adeps Lanæ.

LANOLIN. See Adeps Lanæ.

LANOLIN, SULPHURATED. See Thilanin.

LANTANIN.

An alkaloid, occuring in the herb *Lantana Brasiliensis*. Forms a white, bitter, crystalline powder, which is employed as an antiperiodic and antipyretic. Like quinine, it produces a moderate effect on the circulation, determining a retardation of the chemical phenomena of nutrition and a diminution of temperature. It is superior to salts of quinine, as it is tolerated by the most delicate patients ; in larger doses it is a powerful antiperiodic. Intermittent fevers, though antagonistic to sulphate of quinine, have yielded to the administration of 2 Gm. of lantanine. The dose is from 1 Gm. to 2 during the day, given in pills of 10 cg. each, two being given every two hours.

LEPINE.

An antiseptic mixture, of the following composition:

Mercuric chloride	0.001 Gm.
Carbolic acid	0.100 "
Salicylic acid	0.100 "
Benzoic acid	0.050 "
Calcium chloride	0.050 "
Bromine	0.010 "
Quinine hydrobromate	0.200 "
Chloroform	0.200 "
Distilled water	100.00

LEUKO ALIZARIN. See Anthrarobin.

LIGNOSULFIT.

A liquid side product obtained in the manufacture of cellulose by Kellner's method. It is used in treatment of pulmonary troubles by inhaling the vapors; its active ingredient is sulphurous acid, the irritating properties of which are modified by the presence of aromatics.

LIPANIN.

A mixture of olive oil, with 6% of oleic acid, offered as a substitute for cod-liver oil.

LIQUOR ANTHRACIS.

A solution of 100 parts of coal tar in 200 parts of benzol to which has been added 200 parts of 90% alcohol. The mixture is agitated at 35° C. until a uniform fluid results.

LIQUOR ANTHRACIS SIMPLICIS, AND COMPOSITUS.

An antiseptic preparation of coal-tar, of the consistence of a thin fluid, which, when spread in thin layers, evaporates rapidly. A solution of sulphur, resorcin and salicylic acid in Liquor Anthracis Simplicis constitutes the "Compound Solution." Nothing is known as to the solvent and method of preparation

LIQUOR ANTISEPTICUS-VOLKMANN.

An antiseptic solution supposed to contain alcohol (10), Glycerin (20C) and water (100).

LITHIUM DIURETIN. See Uropherin.

LITHIUM SALTS.

BENZOATE. C_6H_5COO Li. Equal molecular weights of lithium carbonate and salicylic acid are brought together with sufficient water and heated until solution has taken place; the resulting solution of lithium benzoate is evaporated to dryness on a water-bath. This salt occurs as a fine, white powder or in scales, which are soluble in 3 parts of cold water and 10 parts of alcohol. Is employed in the treatment of rheumatism in doses of 0.5 to 1 Gm. (8 to 15 grains).

DITHIOSALICYLATE 1.
$$\begin{array}{l} S\text{-}C_6H_3(OH)COO \text{ Li} \\ S\text{-}C_6H_3(OH)COO \text{ Li} \end{array}$$
Obtained by neutralizing dithio-salicylic acid 1 (q. v.) with lithium carbonate. This is a yellow powder, readily soluble in water and insoluble in alcohol. The therapeutic properties and dose of this salt have not been determined.

DITHIOSALICYLATE 2 is obtained by neutralizing dithiosalicylic acid 2 (q. v.) with lithium carbonate. This salt forms an amorphous powder which is soluble in water and alcohol. Employed in treatment of rheumatism and gout.

FORMATE. HCOO Li+H₂O. Obtained by neutralizing formic acid with lithium carbonate, recrystallizing the resulting salt. It forms colorless needles which are very soluble in water. Employed in rheumatism and gout; dose, about 0.2 Gm. (3 grains).

IODATE. See Iodic Acid.

SALICYLATE. $C_6H_4(OH)COO$ Li. is obtained by neutralizing an equivalent amount of salicylic acid (138 p.) with lithium carbonate (37 p.) in the presence of water containing a little alcohol; the resulting solution should have a slight acid reaction. This salt forms a white, readily soluble, crystalline powder, which is employed in treatment of acute and chronic rheumatism in doses of 0.5 to 1 Gm. (8 to 15 grains).

SULPHOICHTHYOLATE is obtained by neutralizing ichthyol-sulphonic acid with lithium carbonate. It forms a black looking, tarry-like mass, which is dissolved by water, forming a turbid solution. Employed internally in treatment of rheumatism, in doses of 0.5 Gm. (8 grains).

LOBELINE SULPHATE.

From the leaves of the *Lobelia inflata*. Yellowish white, rather hygroscopic powder. For bronchitis, dyspnœa, and spasmodic forms of asthma. Dose 1 to 6 grains either internally or hypodermically.

LORETIN. See under Chinolin.

LOSOPHAN. $C_6HI_3(OH)(CH_3)$

Synonym: Tri-Iodo-Meta-Cresol.

This is prepared by the action of iodine, in the presence of an alkali, on o-oxy-p-toluic acid. Losophan forms colorless, inodorous crystals, insoluble in water, soluble in ether and the fixed oils, melts at 121.5° C. (250.5° F.), and contains 78.4 per cent of iodine. It is employed in various parasitic affections of the skin in alcoholic solution (1 to 2%), or as an ointment (1 to 10%).

LUCILLINE.

A pure petroleum jelly.

LUPERINE.

A mixture of powdered gentian, columbo and quassia. Remedy against dipsomania.

LYCETOL. (Di-methyl-piperazine tartrate) $(NH(CH_2.CHCH_3)_2NH)$.

This is obtained by distilling glycerin with ammonium bromide, and reducing the resulting product (di-methyl-pyrazine) with metallic sodium; is said to be more efficacious as a solvent for uric acid than piperazine. The dose is the same as that of piperazine.

LYSIDIN. $(CH_2N)(CH_2NH)C.CH_3$.

Synonym: Methyl-glyoxalidin.

This base is obtained by the interaction between ethylen-diamine hydrochloride and sodium acetate, liberating the base from its salt by means of a caustic alkali. It is described as being a bright red-colored crystalline mass, very hygroscopic and characterized by a peculiar mouse-like odor. Because of its extremely hygroscopic nature it is now placed on the market in the form of a 50 per cent. solution. This is a pale yellowish liquid of soap-like feeling when rubbed between the fingers; its sp. gr. is 1.054 and boiling point about 105° C. (226.4° F.). It precipitates solutions of mercuric chloride and iodide, soluble in excess of lysidin; ferric chloride forms a brown precipitate, soluble in excess of the reagent. One gramme of lysidin (cryst.) requires 5 Cc. of $\frac{N}{1}$ hydrochloric acid v.s. for neutralization, phenol-phtalein being the indicator. Lysidin is recommended as a solvent for uric acid deposits, being given in doses of 1 to 5 Gm. (15 to 75 grains) daily, dissolved i excess of carbonated water. Where the solution (50%) is employed au equivale double amount i used.

LYSIDIN-BITARTRATE.

A soluble, white, crystalline powder, 10 Gm. of which correspond to 7.2 Gm. of the 50% solution, or 3.6 Gm. of pure lysidin.

LYSOL. See under Cresol.

LYSOLUM BOHEMICUM.

A dark-brown liquid of agreeable odor, which is miscible with water in all proportions, forming transparent yellow solutions. One to 2 per cent. solutions are used as a disinfecting wash, while a 0.2 per cent. solution is sufficiently strong for washing instruments.

MAGNESIUM SALTS.

BENZOATE. $(C_6H_5COO)_2Mg$. Magnesium carbonate is mixed with sufficient water to form a smooth paste; to this is then added an equivalent (molecular) quantity of benzoic acid, and the solution evaporated to dryness. It forms a white crystalline powder, soluble in 20 parts of cold water. Employed in the treatment of gout, urinary calculi, etc. It has been recommended by Klebs in treatment of tuberculosis.

LACTATE. $(C_3H_5O_3)_2Mg + 3H_2O$. Lactic acid, previously diluted with water, is neutralized with magnesium carbonate, evaporated and crystallized. It forms colorless crystals which are soluble in 30 parts of cold water. Employed as a laxative in doses of 1 to 3 Gm. (15 to 45 grains).

PHENOLSULPHONATE. This salt, which occurs as white, almost odorless, needles, with a bitter, not disagreeable, taste, has been recommended as an antiseptic purgative in doses of 15 to 30 grains. The alkaline character of the salt is an advantage, as it diminishes the danger of intestinal irritation. The salt is soluble in 2 parts of water and 5 parts of alcohol.

SALICYLATE. $(C_6H_4(OH)COO)_2Mg + 4H_2O$. Salicylic acid is dissolved in boiling water and neutralized with magnesium carbonate, evaporated and crystallized. It forms colorless, hygroscopic crystals which are readily soluble in water and alcohol. Employed in abdominal typhus in doses of 1 to 2 Gm. (15 to 30 grains).

SULPHOPHENATE. $(C_6H_5SO_4)_2Mg$. Occurs in white needles, soluble in **water and alcohol**. Recommended as a laxative and intestinal antiseptic in doses of **1 to 2 Gm. (15 to 30 grains)**.

MALAKIN. $C_6H_4(OC_2H_5).N:CH.C_6H_4(OH)$.

Synonyms; Salicyl-phenetidine ; Salicyliden-pnenetidine.

This is a condensation product of *p*-phenetidine and salicylic aldehyde, occurring in bright yellow needles, which melt at 92° C. (197.6° F.), almost insoluble in water and alcohol and decomposed by dilute mineral acids. Malakin is recommended as an antipyretic and and antirheumatic in doses of 1 Gm. (15 grains).

MALLEIN. (Dry.)

An analogue of tuberculin, employed as a diagnostic of glanders. It is said to be a mixture of the poisonous, active metabolic products of the bacillus of glanders, and to be obtained rom extremely virulent cultures of this bacillus; it is said to be stable and can be kept for years. Mallein occurs as a yellowish-white bulky powder, readily soluble in water ; from 4 to 5 Cg (3-5 to 4-5 grains) are dissolved in freshly boiled distilled water and injected with strict antiseptic precautions, into the middle of the side of the neck.

MANGANESE ALBUMINATE.

Yellowish white scales, soluble in water.

MANGANESE SACCHARATE.

Brown powder soluble in water. The albuminate, peptonate and saccharate have been recommended in the same way as hæmogallol.

MARROL.

A new dietetic preparation, said to consist of ox marrow, malt extract and hop extract.

MATZOON. See Kefir.

MECONARCEINE.

A derivative of narceine, which appears in lemon-yellow crystals melting at 126° C. (358.8° F.), soluble in dilute alcohol, but very slightly in water. Recommended in bronchial affections, neuralgias and insomnia, in doses of 0.01 to 0.03 Gm. (1-6 to ⅙ grain).

MEDULLADEN.

An extract prepared from bone marc. Recommended in treatment of gout ; also anæmia.

MEDULLARY GLYCERIDE.

Glycerin extracts of bone marrow from calves' ribs. Tonic. Anæmia.

MELANTHINE.

A glucoside obtained from the seed of *Nigella sativa*, resembling in character sapotoxin (obtained from quillaya bark). It is regarded by Kobert and Schulz as one of the series of saponines, whose typical physiological properties it possesses being however, considerably more toxic than others of the series

MENTHENE.

A liquid extracted from menthol, of the composition $C_{10}H_{18}$, soluble in **alcohol**, ether and benzin, and insoluble in water. It boils at 163° C., and has the density, at zero, C., of 0.8326.

MERCURO-IODO-HÆMOL. See under Hæmol.

MERCURY SALTS AND COMPOUNDS.

MERCUROUS ACETATE. $(CH_3COO)_2Hg_2$. A solution of mercurous nitrate is poured, under constant stirring and away from access of light, into a cold solution of sodium acetate; the precipitate formed is allowed to stand in a cool place for 12

hours, then washed with a little water and alcohol, and dried at a low temperature, it forms white, glassy scales, which turn gray on exposure to heat or light, particularly when moistened with water; soluble in 300 parts of water and insoluble in alcohol. Employed in treatment of syphilis in form of pill, dose being 0.01 to 0.06 Gm. (1-6 to 1 grain), externally in ointments (1:10:25).

MERCURY ÆTHYLCHLORIDE is obtained by mixing equal parts of mercuric chloride (dissolved in alcohol) and mercury ethide; the resulting æthylchloride forms colorless shining scales, of unpleasant ethereal odor, slightly soluble in water and alcohol. Because of its indifference to albumen, it is recommended for subcutaneous injection in place of mercuric chloride.

MERCURY AMIDOPROPIONATE, OR ALANATE, is prepared by neutralizing amidopropionic acid with mercuric oxide, evaporating and crystallizing. Forms a white crystalline soluble powder. Employed for subcutaneous injections in place of mercuric chloride. •
Dose, 0.005 to 0.01 Gm. (1-12 to 1-6 grain).

MERCURY ALBUMINATE is obtained by pouring a solution of albumen (1:8) into a 4% mercuric chloride solution, the former being in slight excess. The solution is allowed to stand for 48 hours; the clear solution is then decanted from the precipitate, which is at once mixed with sugar of milk and dried in an exsiccator, adding sufficient milk sugar, so that the resulting powder contains 1 to 1.5% of mercury albuminate. This preparation is used as an antiseptic dusting powder in surgery.

MERCURIC-OXIDE-ASPARAGIN, $(C_2H_3(NH_2)(CONH_2)(COO)_2Hg.$ is an aqueous solution prepared by adding 0.72 Gm. of freshly precipitated mercuric oxide to a solution of 1 gramme of asparagin in 5 Gm. of water, shaking frequently for some time, filtering, and adding water to 72 Cc. This solution (1%) is used subcutaneously in treating syphilitic diseases.

MERCURIC BENZOATE. $(C_6H_5COO)_2Hg+H_2O,$ is obtained by precipitating a solution of mercuric nitrate with a solution of sodium benzoate. This forms a white, crystalline, inodorous, tasteless powder, slightly soluble in water, but readily soluble in a solution of common salt. Employed subcutaneously in treatment of syphilis, the solution being prepared from 3 parts of the benzoate, 1 part of sodium chloride and 400 parts of water, one syringeful being given daily.

MERCURIC-CHLORIDE-UREA. To a cold solution of 1 Gm. of mercuric chloride in 100 Cc. of water, 0.5 Gm. of urea is added, and, when solution has taken place, filtered. Employed subcutaneously in syphilis; 1 Cc. of the solution contains 0.01 Gm. of mercuric chloride.

MERCURIC CYANIDE, $Hg(CN)_2,$ is made by passing hydrocyanic-acid gas through water which contains freshly precipitated yellow oxide of mercury; the solution is then filtered, evaporated and crystallized (caution!!). This forms colorless crystals, which are very soluble in water and alcohol. Employed in syphilitic diseases subcutaneously (0.1 Gm. in 10 Cc. water), ½ to 1 syringeful daily. Great caution should be observed in administering this remedy.

MERCURY CARBOLATE OR PHENATE. $(C_6H_5O)_2Hg+H_2O.$ An alcoholic solution of mercuric chloride is added to an alcoholic solution of sodium phenate, the solution is evaporated to dryness, and the product washed with water, then crystallized from alcohol. Mercury phenate forms colorless needles, almost insoluble in cold water and alcohol. Employed in syphilis in doses of 0.016 to 0.032 Gm. (¼ to ½ grain).

MERCURIC FORMAMIDATE. $(HCONH)_2Hg,$ is a solution resulting from the solvent action of formamide on freshly precipitated mercuric oxide. Each cubic centimeter corresponds to 0.01 Gm. mercuric chloride. Employed subcutaneously in syphilis (0.1 to 10).

MERCURY GALLATE. Molecular quantities of gallic acid and yellow mercuric oxide are mixed with water and evaporated to dryness. This forms a greenish-black, insoluble powder, which is used as an antisyphilitic in place of the less stable tannate.

MERCURO-IODO-HÆMOL. See under Hæmol.

MERCURY IMIDO-SUCCINATE OR ASPARAGINATE. $[C_2H_4(CO)_2N]_2Hg.$ Freshly precipitated mercuric oxide is warmed with an aqueous solution of succinimide, the filtered solution is evaporated and crystallized. It forms a lustrous crystalline powder, soluble in 25 parts of water and 300 parts of alcohol. Subcutaneously injected in doses of 0.01 Gm. (1-6 grain).

MERCURIC IODATE. See Iodic Acid.

MERCURIC NAPHTOLATE. $(C_{10}H_7O)_2Hg.$ A solution of mercuric nitrate is precipitated by means of δ-sodium naphtolate; the resulting precipitate forms, when washed and dried, an inodorous, insoluble powder, employed externally in skin diseases, internally in treatment of typhus. Dose, 0.06 Gm. (1 grain).

MERCURIC OXYCYANIDE, $Hg_2O(CN)_2$, is a white crystalline soluble powder, which is said to be six times more powerful as an antiseptic than corrosive sublimate, at the same time possessing the advantages of being neutral, less caustic and not coagulating albumen. As an antiseptic wash, solutions of 1 to 1,500 are employed

MERCURIC PEPTONATE, a yellowish solution, which contains mercuric chloride 1 part, peptone 3 parts and sodium chloride 3 parts, dissolved in 100 parts of water. This is employed for subcutaneous injections, since it does not cause pain nor produce abscesses. The dose is 1 Cc. corresponding to 0.01 Gm. (1-6 grain) of mercuric chloride.

GLUTINE-PEPTONE-SUBLIMATE is a compound of mercuric chloride (25%) with glutine peptone hydrochloride. This forms a white hygroscopic powder, which is used in subcutaneous injections, the dose being 0.01 Gm. (1-6 grain).

MERCURY-POTASSIUM-HYPOSULPHITE, $3Hg(S_2O_3)_2 + 5K_2S_2O_3$, is prepared by dissolving freshly precipitated mercuric oxide in a solution of potassium hyposulphite, evaporating and crystallizing. This salt is employed for subcutaneous injections.

MERCURIC RESORCINATE. The precipitate obtained by interaction between solutions of mercuric acetate and sodium resorcinate is dissolved in excess of mercuric acetate, evaporated and crystallized. This forms a dark yellow, crystalline powder, insoluble in the usual solvents. For subcutaneous injections the following formula **may** be used:

Hydrargyri Resorc.-acetici - - 5.6 Gm.; Paraffin liquid - - 5.5 Gm.; Lanolin anhyd. 2.0 Gm.; each 1 Cc. contains 0.387 Gm. of mercury; the fluid should be warmed before use, and not more than 0.2 Cc. employed weekly.

MERCURIC SALICYLATE. $C_6H_4 <^O_{COO}> Hg$. This may be prepared by interaction between solutions of sodium salicylate and mercuric nitrate, or by warming together equivalent quantities of salicylic acid and freshly precipitated mercuric oxide in the presence of water on a water-bath, until the yellow mercuric oxide has been entirely converted into the white salicylate. This forms a white, inodorous, tasteless and amorphous powder, which is insoluble in water and alcohol, but is readily dissolved by a solution of sodium chloride or **any** of the halogen salts. It is given in doses 0.001 to 0.008 Gm. (1-64 to ⅙ grain).

MERCURY SOZOIODOL, $(C_6H_2I_2(OH)SO_3)Hg$, is obtained as a precipitate, on mixing concentrated aqueous solutions of sodium sozoiodol and mercuric nitrate. It forms a fine, yellow powder, which is soluble in 500 parts of water, but freely taken up by a solution of sodium chloride or any of the halogen salts. This salt is employed chiefly in the treatment of syphilis, locally and subcutaneously. The subcutaneous dose is 0.06 Gm. (1 grain). For antiseptic applications a 1 to 2% ointment, dusting powder or wash may be employed.

MERCUROUS TANNATE is prepared by rubbing a concentrated solution of mercurous nitrate with a solution of tannin until a pasty mass separates; this is then washed with water by trituration and dried at 40° C. (104° F.). It forms brownish-green scales, which are not soluble (without decomposition) in water or alcohol. Employed in syphilis in doses of 0.06 to 0.13 Gm. (1 to 2 grains).

MERCURY THYMOLATE, $C_{10}H_{13}O$ Hg OH, is obtained by precipitating a solution of mercuric nitrate with sodium-thymolate. An unstable, violet-green powder.

MERCURY THYMOLACETATE, $C_{10}H_{13}O$ Hg+Hg CH_3COO, is prepared by dissolving the above mercury thymolate in a concentrated hot solution of mercuric acetate; on cooling, the double salt crystallizes out. It forms a crystalline insoluble powder, containing 57% of mercury. Both of these salts are used internally in doses of 0.005 to 0.01 Gm. (1-12 to 1-6 grain); when used subcutaneously it should be suspended in paraffin oil as directed under mercuric resorcinate.

MERCURY TRIBROM-PHENOL-ACETATE. A hot solution of tribrom-phenolate of sodium is precipitated by a solution of mercuric acetate, the resulting precipitate is then dissolved in a hot concentrated solution of mercuric acetate, which, on cooling, deposits yellow crystals of the above salt. This contains 29.3% of mercury, and is employed subcutaneously in treatment of syphilis.

META-AMIDO-PHENYL-PARA-METHOXY-CHINOLIN.

Recommended as an antiperiodic and antipyretic in doses of 0.2 to 0.5 Gm. (4 to 8 grains.)

META-CRESOL. $C_6H_4(CH_3)OH.(1-3)$.

One of three isomers which are found in coal and beech-wood tar. This product possesses a very feeble odor, boils at 198° C. and dissolves in 50 parts of water. It is less toxic than carbolic acid and is said to be five times more active in antiseptic power

META-DI-HYDROXY-BENZENE. See Resorcin.

METHACETINE. $C_6H_4.OCH_3.NHCH_3CO.$

Synonyms: Para-acetanisidine; Para-oxymethyl-acetanilide.

This is an analogous compound to phenacetine, being the methyl ester of para phenetidine, or it may be regarded as acetanilide in which a hydrogen in the benzene nucleus is substituted by an oxy-methyl group (-OCH₃). Methacetine forms colorless, inodorous and tasteless, scaly crystals, which melt at 127° C. (260.6° F.), almost insoluble in cold water, but readily soluble in hot water, also in alcohol, glycerin and the fatty oils. It is recommended as an antipyretic and antineuralgic in doses of 0.032 to 0.5 Gm. (½ to 8 grains); 0.5 Gm. of methacetine corresponds to 1 Gm. of phenacetine as antipyretic, and 1 Gm. of the former corresponds to 2 Gm. of the latter as anti-neuralgic.

METHOXY-ANTIPYRIN. (Para.) See under Antipyrin.

METH-OXY-CAFFEINE. $C_8H_{10}(OCH_3)N_4O_2.$

This derivative of caffeine appears in white needles or as an amorphous powder, melting at 177° C. (350.6° F.). Recommended in migraine and neuralgias in doses of 0.26 Gm. (4 grains), hypodermically as a local anæsthetic.

METHYL-ACETANILID. See Exalgine.

METHYLAL. $CH_2(OCH_3)_2.$

Synonym: Methylen-dimethyl-ether.

This fluid is obtained by the abstraction of one molecule of water from a compound of one molecule of formaldehyde and two of methyl-alcohol ; the resulting product belongs to the group of organic bodies known as "acetals." It is a colorless liquid, of an ethereal odor, soluble in water, alcohol and ether ; its sp. gr. is 0.855 (15° C.) and boiling-point is 42° C. (107.6° F.). Methylal is recommended as a hypnotic in doses of 1 to 5 Gm. (15 to 75 grains).

METHYL CHLORIDE $CH_3Cl.$

Synonyms: Chlormethyl; Monochlormethane.

This gaseous compound is made by heating methyl-alcohol and hydrochloric acid under pressure at 100°C.; the gas produced is washed and dried, and then compressed in copper or steel cylinders at low temperature with pressure. Methyl chloride forms a colorless and inflammable gas of an ethereal odor, which, under a pressure of five atmospheres or at a temperature of -25°C., is converted into a liquid. It appears in commerce in the compressed liquid form, which is employed as a spray to produce local anæsthesia. A minute stream of the liquid is directed upon a tampon of wool and silk placed over the surface to be anæsthetized ; the rapid evaporation produced absorbs the heat from the parts and leaves them bloodless and insensible.

RICHARDSON'S COMPOUND LIQUID consists of a mixture of ether and chloroform saturated with methyl chloride ; has been recommended as a substitute for chloroform.

METHYLENE BICHLORIDE. See Methylene Chloride.

METHYLENE BLUE.

Synonym: Tetra-methyl-thionine chloride.

A complex derivative of diphenylamine, classed as an "aniline dye." This salt occurs in dark blue or reddish-brown crystals or crystalline powder, of a bronze-like tinge, slightly soluble in water and alcohol, producing a deep blue solution. Methylene blue is employed internally as an analgesic in neuralgic and rheumatic affections in doses of 0.13 to 0.5 Gm. (2 to 8 grains), or subcutaneously in doses of 0.016 to 0.06 Gm. (¼ to 1 grain).

METHYLENE CHLORIDE. $CH_2Cl_2.$

Synonyms: Dichlormethane ; Methylene bichloride.

Prepared by the reducing action of nascent hydrogen (from zinc and hydrochloric acid) upon chloroform, the product being washed and rectified. This forms a colorless liquid, which resembles chloroform in odor and solubility ; specific gravity is 1.364 (15° C.); boiling point 41.6° C. (107° F.). Like pure chloroform it is readily decomposed by the action of sunlight, hence the addition of one per cent. of alcohol is recommended. Methylene chloride is recommended as an anæsthetic in place of chloroform.

METHYL-GLYOXALIDIN. See Lysidin.

METHYL-PHENACETIN. See under Phenacetine.

METHYL-PROTOCATECHUIC ALDEHYDE. See Vanillin.

METHYLPYRIDENE SULPHOCYANATE.

A compound obtained by the action of thiocyanic acid on quinoline. A crystalline energetic antiseptic, free from caustic properties and all dangers that are liable to occur with the use of corrosive sublimate and carbolic acid. Usually employed in 1 per cent. solutions.

METHYL-VIOLET. See Pyoktanin, blue.

MICROCIDIN. $C_{10}H_7ONa$.

This is a sodium b-naphthol, obtained by fusing b-naphthol with one-half of its weight of sodium hydrate. As an antiseptic wash a ½% aqueous solution is employed. It is recommended internally as an antipyretic, also as an antisepsic in purulent otitis media, rhinitis, ozœna, and tonsillitis. In the first-named affection a 0.3 to 0.4 per cent. solution is used ; in nasal and throat diseases, a 0.1 per cent. solution.

MIGRANIN. See under Antipyrine.

MONOCHLORAL-ANTIPYRINE. See under Antipyrine.

MONO-CHLOR-METHANE See Methyl Chloride.

MONO-PHENETHYDIN. See Apolysine.

MORPHINE SALTS. $C_{17}H_{17}(OH)_2.NO.A.$

Among the newer salts of this base are the :

BENZOATE, obtained by neutralizing morphine with benzoic acid. This forms white, crystalline prisms or powder, which is employed in treatment of asthma in the same doses as the morphine sulphate.

BORATE is recommended for subcutaneous injections and eye washes, because of the stability and neutral nature of the salt.

PHTALATE and *TARTRATE* are recommended for subcutaneous injections, both being very soluble in water.

MORRHUIN. $C_{19}H_{27}N_2$.

A basic principle found in cod liver oil. It is a thick, oily liquid, which is soluble in alcohol and ether ; 2 Mg. of the principle are presumed to represent a tablespoonful of cod liver oil in activity.

MUSCARINE. $(CH_2)_3 - N.OH - C_2H_5O_2$.

An alkaloid obtained from the fungus *Agaricus muscarius* It appears in hygroscopic crystalline masses, readily soluble in water and alcohol. The nitrate and sulphate are usually employed. Used in place of eserine as an antidote to atropine, also recommended for diabetes insipidus. Dose, 0.0022 to 0.0044 Gm. (1-30 to 1-15 grain).

MUSIN.
A fluid preparation made from tamarinds.

MYDRINE.
A white soluble powder, representing a combination of the two mydriatic alkaloids ephedrine and homatropine. Recommended as a mydriatic in solution of 0.3 Gm. (4.5 grains) in 3 Gm. (45.5 grains) of distilled water.

MYRONIN.
The firm of Eggert & Hæckel, in Berlin, have introduced under this name a new ointment base, a vehicle consisting of a mixture of soap, carnauba wax and dœgling oil. The latter possesses remarkable penetrating powers and does not readily become rancid Myronin is prepared by heating stearic acid, in the presence of carnauba wax, with sufficient dilute potassium carbonate solution until saturation has taken place The mixture of the resulting stearin soap and wax is then diluted with the dœgling oil until an ointment-like mass results, possessing the desired degree of consistency. This base, as prepared, contains about 12.5% of water, which may be increased or decreased as desired.

MYRRHOLIN
A solution of equal parts of myrrh and oil (fatty), which has been used as a vehicle for creosote in laryngeal and pulmonary tuberculosis.

MYRTOL.

A mixture of dextro-pinene ($C_{10}H_{16}$), eucalyptol ($C_{10}H_{18}O$) and a camphor-like body ($C_{10}H_{18}O$), obtained by the fractional distillation of the oil of *Myrtus Communis*. It forms a colorless liquid, of aromatic odor, boiling between 160° to 170° C. (320° to 338° F.) ; recommended as a disinfectant and deodorant in putrid bronchitis and other diseases of the respiratory tract. Dose, 5 minims every two or three hours, taken in capsules.

NAP ELLINE.

From *Aconitum napellus*, white powder, soluble in water, alcohol and ether. Anodyne, analgesic. (Neuralgia, rheumatism, etc.) Dose, ⅓ to ½ grain.

NAPHTHALOL. See BetoL

NAPHTHALENE (NAPHTALIN U. S. P.). $C_{10}H_8$.

Synonym : Tar Camphor.

A hydrocarbon obtained from coal tar, which occurs in white scales of fatty lustre and strong coal-tar odor, melts at 80° C. (176° F.) and is soluble in alcohol and ether, but insoluble in water. Used internally as a vermifuge against oxyuris vermicularis, as an expectorant, as an antiseptic in chronic diarrhoea and typhoid fever, also as an antipyretic. Externally naphtalene is used in various skin diseases as eczema, psoriasis, lepra, etc. Dose is 0.13 to 0.5 to 1 Gm. (2 to 8 to 15 grains) , for tapeworms, 1 Gm. followed by castor oil. Externally, from 5 to 10% dusting powder or ointment.

NAPHTHOL (ALPHA). $C_{10}H_7OH$.

A constituent of coal tar, also obtained artificially from Naphthalene. It forms colorless prisms of phenol-like odor and burning taste, soluble in alcohol and ether, slightly so in water, melts at 94° . Alpha Naphtol is an antiseptic and antiferment, being recommended in treatment of diarrhoea, dysentery, typhoid fever, etc. A solution of 0.1 to 0.25p. in 1,000 prevents the development of the spores of the tubercle bacilli. Said to be 1½ times as strong as beta-naphtol.

NAPHTHOL (BETA). $C_{10}H_7OH$.

A constituent of coal tar, also obtained artificially from naphtalene. For description and tests see U. S. Pharmacopœia. Beta-naphtol is used as a general antiseptic in cutaneous disorders and in affections of the respiratory tract, also as an intestinal antiseptic in typhoid and typhus fevers and in chronic diarrhoeas. The dose varies from 0.12 to 1 Gm. (2 to 15 grains). Externally, 2 to 10 per cent. solutions or ointments are employed.

β-NAPHTHOL-ANTIPYRINE. See Naphthopyrin.

NAPHTHOL-ARISTOL. (Di-iod-b-naphthol).

Prepared like aristol ; a solution of sodium naphthol (b-naphthol 11p. and sodium carbonate 4p.) is precipitated by an aqueous solution of iodine in potassium iodide (2.4p. each). It is a greenish-yellow powder, insoluble in water, slightly soluble in alcohol and very soluble in chloroform. The compound is recommended as an antiseptic.

β-NAPHTHOL BENZOATE. See Benzonaphthol.

β-NAPHTHOL CAMPHOR.

Synonym : Naphthol, camphorated.

A syrupy fluid made by fusing together naphthol 1 part and camphor 2 parts. This is used as an antiseptic application to boils and in tuberculosis.

β-NAPHTHOL CARBONATE. $CO(OC_{10}H_7)_2$.

A di-naphthyl ester of carbonic acid. It is prepared by the action of phosgene on b-naphthol sodium, yielding shining, colorless scales, which are insoluble in water and melt at 176° C. (348.8° F.). Recommended as a substitute for b-naphthol, as an intestinal antiseptic, owing to its less irritating qualities.

NAPHTHOL CARBOXYLIC ACID. $C_{10}H_6(OH)CO_2H$.

Synonym : x-Oxynaphtholc Acid.

This is obtained by the action of carbonic acid gas upon sodium-x-naphthol under pressure. It forms a white crystalline powder or acicular crystals ; melts at 186° C. (366.8° F.) ; insoluble in water, soluble in alcohol, ether, fatty oils and glycerin. Forms soluble salts with the alkalies or alkali carbonates Recommended as an antiseptic disinfectant and antiparasitic, in form of an ointment (5 to 10%) or antiseptic gauze (1%).

β-NAPHTHOL-DISULPHONATE OF ALUMINUM. See Alumnol.

β-NAPHTHOL-x-MONO-SULPHONATE OF CALCIUM. See Asaprol.

NAPHTHOPYRIN.

Synonym: b-Naphthol-antipyrine.

Made by triturating together equal molecular quantities of b-Naphthol and antipyrin, forming thereby a tough mass, which gradually assumes a crystalline form ; insoluble in water and soluble in alcohol.

NAPHTHOPYRINE. See under Antipyrine.

NAPHTHOSALOL. See Betol.

NARCEINE HYDROCHLORATE. $C_{23}H_{29}NO_9.HCl+3H_2O.$

Narceine is found in opium to the extent of 0.1 to 0.4%. The hydrochlorate forms colorless, crystalline needles, which are soluble in water and alcohol. It is employed as a hypnotic in doses of 0.01 to 0.06 Gm. (1-6 to 1 grain).

NARCOTIN. $C_{22}H_{23}NO_7.$

This alkaloid, which is found in opium, occurs in colorless crystals, which melt at 176° C. (348.8° F.) ; insoluble in water and alkalies, but readily soluble in alcohol and ether. It is used as a hypnotic in doses of 0.25 to 1 Gm. (3.8 to 15 grains).

NASROL. See Sodium Sulfocaffeate.

NEURINE. $N(CH_3)(C_2H_3)OH.$

Synonym: Tri-methyl-vinyl-ammonium-hydroxide.

This base, containing the unsaturated radicle "vinyl," C_2H_3, is found along with neuridine ($C_5H_{14}N_2$) among the products of the decomposition of flesh. Obtained synthetically by reaction between æthylene bromide and alcoholic trimethylamine at 60° C. (140° F.) under pressure. It forms a very poisonous, strongly alkaline fluid, which is very soluble in water, but is decomposed on boiling. A 3% solution is employed as a local application for diphtheritic membranes.

NEURODINE. $C_6H_4(OCO.CH_3) NH.COOC_2H_5.$

Synonym: Acetyl-p-ethoxy-phenyl-urethane.

A substance introduced by Merck as an antipyretic and antineuralgic. By the action of chloro-carbonic ether (CO(Cl) (OC_2H_5)), an amido-phenol ($C_6H_4(OH)NH_2$), para-oxyphenyl-urethane, is formed, which on being acetylated is converted into neurodine. This forms colorless, inodorous crystals, melting at 87° C. (188.6° F.), soluble in 1400p. of cold water.

Dose as antineuralgic is 1 to 1.5 Gm. (15 to 23 grains), as antipyretic 0.5 Gm. (7½ grains).

NEUROSIN.

A French preparation (in syrup or granule form) which contains as active constituent glycerino-calcium phosphate.

NITRO-SALOL. See under Salol.

NOSOPHEN.

Synonym: Tetra-iodo-phenol-phtalein ($C_6H_2I_2.OH)_2$. $C<^{C_6H_4CO}_{O--}$

Obtained by the action of iodine on a solution of phenol-phtalein. It forms a pale yellow colored, inodorous and tasteless powder, insoluble in water and alcohol, melts at 250° C. With alkalies nosophen forms soluble salts, the sodium compound being blue colored. It contains 60 per cent. of iodine, which is not liberated by the action of alkalies or boiling dilute acids. Employed as antiseptic dusting-powder, being destructive to bacteria life.

NUCLEIN.

A phosphorated proteid extracted from the spleen and other organs. It forms a pale yellow-colored powder, soluble in alkaline solutions, but insoluble in alcohol or water. In doses of 2 to 3 Gm. well diluted nuclein is said to enhance phagocytosis by increasing the number of white corpuscles. Also recommended hypodermically in treatment of pleurisy and pneumonia.

NUTRIN.

A dietetic food which according to its manufacturers, represents "the pure nutritious substance of meat."

NUTROL.

An artificially digested starch, containing a small quantity of hydrochloric acid and digestive ferments.

ODONTODOL.

Said to be a mixture of Cocaine Hydrochlorate (1), Oil of Cherry Laurel (1), Tincture Arnica (10), and solution of Ammonium Acetate (20). Used as a dental anodyne.

ŒNANTHOTOXINE. $C_{17}H_{22}O_5$.

A resinous substance obtained from *Œnanthe crocata*; it is said to be very poisonous, and to produce violent spasms as picrotoxin does.

OLEO-CREOSOTE. See under Guaiacol.

ORCHIDIN.

Prof. Poehl's Testicular Fluid. Recommended as a nervine.

OREXIN. $C_6H_4NCH.CH_2NC_6H_5$.

Synonym: Phenyl-dihydro-chin-azoline.

This, a complex chinoline derivative, is obtained by the action of sodium formanilid on o-nitrobenzyl chloride, the resulting o-nitrobenzylformanilid is reduced to the corresponding amido derivative by means of nascent hydrogen; the hydrochlorate of this base on heating gives up one molecule of water, yielding the hydrochlorate of orexin, from which the base orexin may be obtained by treatment with an alkali.

This occurs as a white, amorphous, tasteless powder, which is almost insoluble in water. Employed as a stomachic, stimulating the appetite, also as an antiemetic, given in doses of 0.13 to 0.4 Gm. (2 to 6 grains) either in wafers or capsules, followed by draughts of beef tea or cocoa to prevent any local irritation.

OREXIN HYDROCHLORIDE. $C_6H_4.NCH\ CH_2NC_6H_5.HCl+2H_2O$.

Forms colorless, odorless crystals, of bitter pungent taste, melting at 80° C. (176 °F), soluble in 15 parts of cold water. The properties and dose are same as those of Orexin.

ORTHIN. $C_6H_3.OH.COOH.NH\ NH_2$.

Synonym: Ortho-hydrazine-para-oxybenzoate.

Orthin is a derivative of phenyl-hydrazine $(C_6H_5-NH-NH_2)$ in which one hydrogen atom of the benzene nucleus is replaced by a hydroxyl-group and another by the carboxyl group. The *hydrochlorate*, in which form it usually appears, forms colorless soluble crystals, and is recommended as an antipyretic. Experiments have shown it to be too dangerous for general use.

ORTHO-AMIDO-SALICYLIC ACID. $C_6H_3(NH_2)$ (OH) COOH.

Obtained through reduction of ortho-nitrosalicylic acid. It forms a gray-white, amorphous, inodorous powder, of a faint sweet taste. Insoluble in water, alcohol and ether.

It is employed in treatment of chronic rheumatism, in doses of C.25 to 0.5 Gm. (3 to 7 grains).

ORTHO-HYDRAZINE-PARA-OXY-BENZOATE. See Orthin.

ORTHO-OXY-BENZOIC ACID.

Synonym: Salicylic Acid.

ORTHO-PHENOL-SULPHONIC ACID. See Aseptol.

ORTHO-PHENOLSULPHURIC ACID. See Aseptol.

ORTHO-SULPHAMINE-BENZOIC-ANHYDRIDE. See Saccharin.

ORTHO-SULPHO-CARBOLIC ACID. See Aseptol.

ORTHO-TOLYL-ACETAMIDE. See Aceto-ortho-toluid.

ORTHO-TOLYL-DI-METHYL-PYRAZOLON. See Tolypyrin.

OSMIC ACID. $Os\ O_4$.

Obtained by heating osmium in fine powder in a current of oxygen or moist chlorine; or by evaporating a solution of osmium in nitromuriatic acid to dryness. Occurs in colorless or yellowish-green, lustrous needles, of a penetrating chlorine-like odor, melting at 40° C. (104° F.); stains the skin and linen black. Externally in a 1 per cent. solution (fresh) osmic acid acts as a caustic for cancerous and scrofulous sores; subcutaneously (1% solution) it has been employed in neuralgia, epilepsy, sarcoma, cancer, etc. Dose internally 0.001 Gm. (1-64 grain) in pill form, freshly prepared.

OUABAIN.

A glucosidal principle obtained from the wood of *Acocanthera ouabaio*, also the seed of *Strophanthus glabrus*. Inodorous white crystals, slightly soluble in cold, freely in hot water and dilute alcohol, melting at 200° C. (392° F.). Employed in doses of 0.000065 Gm. (1-1000 grain) repeated at frequent intervals, for the relief of whooping-cough.

x-OXYNAPHTHOIC ACID. $C_{10}H_6(OH).COOH.$

Synonym : *x*-Naphthol-carboxylic acid.

Obtained by the action of carbonic acid on *x*-Naphthol sodium under pressure **at 140° C.** ; the resulting sodium salt is a yellowish powder or colorless needles, of melting-point 186° C. (366.8° F.). Very slightly soluble in cold water, but very soluble in alcohol. The acid is quite soluble in an aqueous solution of borax. Its properties are those of an antiseptic and antizymotic.

Applied **as an ointment (10%) in treatment of scabies.**

OXY-SPARTEINE. $C_{15}H_{24}N_2O.$

This is an oxidation product of sparteine (q. v.), and occurs in colorless, hygroscopic crystals, melting at 81° C. (181.4° F.) and soluble in the usual solvents. It is employed subcutaneously as a heart stimulant ; it, however, lowers the pulse at the same time. Dose is 0.04 Gm. (3-5 grain) gradually increasing to 0.1 Gm. (1.5 grains).

The *hydrochlorate* melts between 48° and 50° C. (118.4° to 122° F.). This alkaloid is incompatible (therapeutically) with opiates.

OZALIN.

A disinfectant, said to consist of a mixture of Calcium, Magnesium and Iron Sulphates, with Caustic Soda and Magnesia.

PANGADUINE.

Name proposed for the collective alkaloids of cod liver oil. A crystalline solid, soluble in alcohol and in a mixture of water and glycerin.

PAPAIN.

Synonyms : Papayotin ; Plant Pepsin ; Papoid.

The concentrated active principle of the juice of the unripe fruit of *Carica papaya*. The juice is concentrated in vacuo, and the ferment is precipitated by the addition of alcohol. Papain forms an amorphous, white, hygroscopic powder, soluble in water and glycerin only. Employed as a digestive ferment ; like animal ferments (pepsin) it digests albuminous substances, possessing the advantage of being active in either acid, alkaline or neutral solutions. Dose 0.12 to 0.3 Gm. (2 to 5 grains) after meals. Applied as a 5 per cent. solution (in equal parts of glycerine and water) it is used to dissolve the false **membranes of** croup and diphtheria.

PAPAYOTIN. See Papain.

PARA-ACET-AMIDO-PHENETOL. See Phenacetine.

PARA-ACET-ANISIDINE. See Methacetine.

PARA-ACETO-PHENOL-ETHYL-CARBONATE.

A **crystalline, colorless** and tasteless powder, soluble in alcohol and insoluble in water. It is recommended as antithermic, analgesic and hypnotic in doses of 0.5 Gm. (8 grains).

PARA-ACET-PHENETIDIN. See Phenacetin.

PARA-ÆTHOXY-PHENYL-URETHANE. See Thermodin.

PARA-ALLYL-PHENYL-METHYL ETHER. See Anethol.

PARA-BROM-ACETANILID. See Antisepsine.

PARA-CHLOR-PHENOL.

A crystalline substitution product of phenol. When fused it congeals at 33° C. ; it is soluble in alcohol, ether and fatty oils, almost insoluble in water. Karpow recommends this as a powerful antiseptic and disinfectant, only exceeded in intensity by silver nitrate and mercuric chloride. It has been successfully employed in treatment of erysipelas in a 2 to 3% vaseline ointment.

PARA-CHLOR-SALOL.

Recommended as superior to salol **as a disinfectant, without possessing any of its** secondary toxic properties. •

PARACOTOIN. $C_9H_{12}O_6.$

A principle which occurs, along with several others, in the Para Coto bark. It forms a pale-yellow, tasteless, crystalline powder, melting at 152° C. (305.6° F.), almost insoluble in water but readily soluble in alcohol. Employed in intestinal catarrh and as an anti-diarrhœic, the dose being 0.1 to 0.2 Gm. (1.5 to 3 grains) for children.

PARA-CRESOL. See under Cresol.

PARA-DIPHENOL. See Hydroquinone.

PARAFORM.

A polymeric formaldehyde, obtained by heating an aqueous solution of formaldehyde, whereby some is volatilized and a polymeric form (paraform) remains behind. It forms a white crystalline powder which is insoluble in water. Recommended as an intestinal antiseptic.

PARA-METHOXY-ANTIPYRIN. See under Antipyrine.

PARA-MONO-CHLOR-PHENOL. See under Chlorphenol.

PARA-OXY-ETHYL-ACETANILIDE. See Phenacetine.

PARA-OXY-METHYL-ACETANILIDE. See Methacetine.

PARA-TOLYL-DIMETHYL-PYRAZOLON. See Tolypyrin.

PASTE, PEPTON. See Pepton-Paste.

PASTE, SERUM. See Serum Paste.

PASTE, SULPHURIC ACID. See Sulphuric Acid Paste.

PELAGIN.

A proprietary remedy for sea sickness, said to be a solution of antipyrine, caffeine and cocaine.

PELLETIERINE, (Punicine). $C_8H_{15}NO$.

One of several alkaloids obtained from the root-bark of the *Punica granatum*. It is a colorless liquid, soluble in 2.5 parts of water, readily in all proportions in alcohol and ether. With acids it unites to form crystalline salts, among which the *tannate* is most frequently employed.

PELLETIERINE TANNATE.

A yellowish, tasteless, amorphous powder, insoluble in water, soluble in 80p. of alcohol, and readily in diluted acids. Prompt and innocuous tænicide in doses of 1.5 Gm. (23 grains) followed by a purgative.

PENTAL. $(CH_3)_2C.CH.CH_3$.

Synonyms Trimethyl-ethylene; b-Isoamylene.

This hydrocarbon is prepared by distilling fusel-oil with zinc chloride, and then treating the distillate (amylene) with concentrated sulphuric acid. Pental is a colorless liquid of sp. gr. 0.6733 (0° C.), boiling at 38° C. (100.4° F), insoluble in water, miscible with alcohol (9%), ether and chloroform in all proportions. Employed as anæsthetic in dental surgery.

PEPSIN, PLANT. See Papain.

PEPTON-PASTE.

Recommended as a substitute for collodion in securing gauzes or bandages to the skin, free from any irritating or contractile properties. The dried varnish-like cuticle may be removed by washing with water.

PEREIRIN.

An alkaloid obtained from the bark of *Geissospermum læve*, by exhausting with boiling alcohol, evaporating to an extract and treating the residue with soda and ether. Forms a white, amorphous powder, slightly soluble in water, readily soluble in alcohol and ether. With acids it yields crystallizable soluble salts; among those usually employed are the *hydrochloride* and *valerianate*. Pereirin is recommended as a tonic and antifebrile, 0.5 to 2 Gm. (8 to 30 grains), being administered about four hours before the expected attack.

PHEDURETIN.

A phenol derivative, composition not given; tasteless white crystals, insoluble in water. Doses of 0.5 to 1 Gm. a powerful diuretic.

PHENACETINE. $C_6H_4.OC_2H_5.NHCOCH_3$.

Synonyms: Para-Acet-Phenetidine ; Para-Acet-Amido-Phenetol ; **Para-Oxyethyl-Acetanilide.** Phenetidine.

This compound, which chemically is closely connected with acetanilide and methacetine, is prepared as follows :

Sodium Para-Nitrophenol, (which is prepared by the action of nitric acid on phenol and subsequent treatment of the product (para) with sodium hydrate), is ethylated by heating with ethyl bromide under pressure ; this product (p-nitrophenetol) is reduced by nascent hydrogen to p-amidophenetol, which by prolonged boiling with glacial acetic acid yields phenacetine. This forms colorless, tasteless, inodorous, scaly crystals,

THE NEWER REMEDIES.   61

melting at 135° C. (275° F.), soluble in 1,500 parts of cold, and 80 parts of boiling water and in about 16 parts of alcohol. Employed as an antipyretic and antineuralgic in doses of 0.5 to 1 Gm. (8 to 15 grains). Decomposed by strong alkalies or acids.

IODOPHENINE, $C_{20}H_{25}I_3N_2O_4$, is a derivative of phenacetine, obtained by the combination of two molecules of the latter with three molecules of iodine. Phenacetine is dissolved in glacial acetic acid, then diluted with water and hydrochloric acid; to this is added an aqueous solution of iodine in potassium iodide until no further precipitation occurs. If the operation is carried on in a hot solution, iodophenine is obtained in brown needle-like crystals, which have an odor resembling that of iodine, melting at 130° to 131° C. (266 to 267.8° F.), soluble in alcohol,and when mixed or heated with water iodine is liberated. This compound contains iodine (51%) in a very loose state of combination; hence it is not adapted for internal use because of its irritating properties.

METHYL-PHENACETINE, $C_6H_4(OC_2H_5)N(CH_3)CH_3CO$, is prepared by the action of methyl iodide on phenacetine-sodium, the latter resulting from the action of metallic sodium on a solution of phenacetine in boiling xylol. This salt forms colorless crystals, melting at 40° C. (104° F.), only slightly soluble in water, but readily so in alcohol. Employed as a hypnotic.

ETHYL-PHENACETINE. $C_6H_4(OC_2H_5)N(C_2H_5)CH_3CO$, a homologue of the above methyl-compound, is prepared in like manner by the action of ethyl iodide on phenacetine-sodium. This forms a yellow colored oil, boiling at 330° to 335° C, (626° to 635° F.), almost insoluble in water, readily so in alcohol and ether. Possesses hypnotic properties, but to a lesser degree than the above methyl derivative.

FORMYL-PHENETIDINE, or *PARA-OXYETHYL-FORMANILID,* C_6H_4 (OC_2H_5)NH.COH, is prepared by heating a mixture of hydrochlorate of p-phenetidine, anhydrous sodium formate and formic acid, the reaction product being crystallized from water. Colorless, inodorous, tasteless crystals, melting at 60°C. (140° F.), slightly soluble in cold water, readily in hot water, alcohol and ether. Recommended as an antiseptic.

LACTOPHENINE, or *LACTYL-PHENETIDINE,* is produced by the action of lactic acid on phenetidine in presence of dehydrating agents. It forms a crystalline powder of bitter taste, more soluble than phenacetine, employed as an antipyretic and sedative in doses of 0.5 to 1 Gm. (8 to 15 grains).

SEDATIN, or *VALERYL-PHENETIDINE,* is obtained by the action of valeric acid on p-amido-phenetol. Recommended as an antipyretic and antineuralgic. The selection of the term "sedatin" is unfortunate, since this was formerly applied as a synonym for antipyrine.

PHENATOL.

Said to be a mixture of acetanilid, caffeine, sodium bicarbonate, carbonate, sulphate and chloride.
Recommended as an antipyretic and anodyne.

PHENAZONE. See Antipyrine.

PHENETIDINE. See under Phenacetine.

β-PHENETIDYL-CROTONIC-ETHYL-ESTER.

This compound is formed by mixing molecular quantities of para-phenetidin (C_6H_4 OC_2H_5NH_2) and aceto-acetic ester. (CH_3.CO.CH_2.CO_2C_2H_5); the mixture becomes turbid and quite warm, with the separation of water. By recrystallization from alcohol, the substance forms glossy white, needle-like crystals, which melt at 53° C, (127.4° F.), insoluble in water, readily soluble in alcohol and ether. Nothing is known concerning the medicinal properties of this substance, as it is still in the experimental stage.

PHENOCOLL HYDROCHLORIDE. $C_6H_4 < ^{OC_2H_5}_{NHCO.CH_2NH_2HCl}$.

Synonyms: Amido-Acet-Para-Phenetidine-Hydrochloride ; Glycocoll-Para-Phenetidine Hydrochloride.

The base phenocoll is prepared by interaction between amido-acetic-acid ester and phenetidine [$C_6H_4(OC_2H_5)(NH_2)$] or by the action of chloracetylchloride on phenetidine, and subsequent treatment of the resulting mono-chlor-acet-para-phenetidine with ammonia. This base unites with acids forming salts. Phenocoll hydrochloride forms colorless needles or a crystalline powder, soluble in 16 parts of water, but the base phenocoll is precipitated upon the addition of alkalies. Because of its greater solubility and freedom from the unpleasant after effects which sometimes accompany phenacetin,it has become quite popular. Employed as an antipyretic, antirheumatic and antineuralgic, in doses of 0.3 to 1 Gm. (5 to 15 grains). Has been recommended as a specific in malarial complaints, also as an antiseptic in treatment of wounds, sores, etc., either in form of powder, 5 per cent. solutions, 10 to 20 per cent., gauze, or ointment (10 to 20%). Decomposed by alkali hydrates or carbonates; also by Piperazin.

PHENOCOLL CARBONATE. [C₆H₄(OC₂H₅)NH.CO–CH₂NH₂]₂CO₂, is a colorless, almost tasteless, crystalline powder, which is quite insoluble in water. When heated to 95°C. carbonic acid is given off. Employed like the above, but particularly adapted as an antiseptic powder.

PHENOCOLL ACETATE forms bulky, crystalline needles which are soluble in 3 parts of water. This salt is adapted for subcutaneous injections.

SALOCOLL, or *PHENOCOLL SALICYLATE*, is a crystalline salt which is less soluble than the hydrochloride, combining the medicinal effects of phenocoll with those of salicylic acid. Employed as an antipyretic and antirheumatic in doses of 1 to 2 Gm. (15 to 30 grains).

PHENOL. (Carbolic Acid.)

PHENOL CAMPHOR.

Synonym: Camphor, Phenylated.

Crystallized carbolic acid is liquefied by heat, and in this is dissolved an equal quantity of camphor; another formula directs three times the quantity of camphor; still another directs 4 parts of camphor, 12 parts of carbolic acid, and 1 part of water. This phenylated camphor is an oily, colorless liquid, which is insoluble in water, very soluble in alcohol, ether, chloroform, fatty and volatile oils. It is employed as an antiseptic, germicide and local anaesthetic (particularly in toothache.)

PHENOL, MONO-BROMO. C₆H₄.Br.OH.

Synonym: Ortho-mono-bromo-phenol.

An oily violet colored liquid, of strong odor, soluble in ether, in 100 parts of water, boils at 195° C. Used as an antiseptic in 1 to 2 per cent. solutions or applications.

PHENOL, MONO-CHLORO (o). C₆H₄ClOH (2:1)

Synonym: Ortho-mono-chloro-phenol.

Obtained by the action of chlorine on amido-phenol. It forms a colorless fluid, soluble in alcohol. Used as an antiseptic, usually with petrolatum in skin diseases.

PHENOL, MONO-CHLORO (p). C₆H₄ClOH. (4:1)

Synonym: Para-mono-chloro-phenol.

This forms crystals soluble in alkalies, ether, slightly in water, melting at 37° C., boiling at 217° C. Properties antiseptic. Used externally in syphilitic diseases of the eyes, tuberculous diseases of the throat in 5 to 20% solution in glycerin. In lupus it is applied heated to 40° C., using afterwards a 25% ointment; after 12 hours this paste is removed and salicylated ointment substituted.

PHENOL AND SODIUM-RICINATE.

A yellowish fluid containing 20% synthetic carbolic acid and 80% sodium sulphoricinate. An antiseptic and rubefacient, used in 20% aqueous solution for painting false diphtheritic membranes, skin diseases, etc.

PHENOLIN. See under Cresol.

PHENOPYRINE. See under Antipyrine.

PHENOSALYL. See under Cresol.

PHENOSUCCIN.

A compound obtained by the action of succinic acid on p-amidophenol. Occurs in colorless needles, which melt at 155°C. (311° F.), insoluble in cold water. Recommended as antipyretic and antineuralgic. The sodium salt (sodium phenosuccinate) is to be preferred.

PHEN-OXY-CAFFEINE. C₈H₉(OC₆H₅)N₄O₂.

A white crystalline powder soluble in alcohol, melting at 142° C. Used as an anaesthetic, antineuralgic and narcotic in doses of 0.26 Gm. (4 grains).

β-PHENYL-ACRYLIC ACID. See Cinnamic Acid.

PHENYL-BORIC ACID. C₆H₅-B(OH)₂.

Prepared by the action of phosphorus oxychloride on a mixture of boric acid and phenol in molecular proportions. It occurs as a white powder, difficultly soluble in cold water. It is employed as an antiseptic, for dressing wounds and ulcers. A one per cent. solution is sufficient to prevent putrefaction.

PHENYL-DIHYDRO-CHINAZOLINE. See Orexin.

PHENYL-HYDRAZINE-LÆVULINIC ACID See Antithermin.

β-PHENYL-PROPIONIC ACID.

Synonym: Hydrocinnamic acid.

Obtained by the reduction of cinnamic acid by means of mercury amalgam. It occurs in colorless crystals, slightly soluble in cold, but very soluble in hot water and alcohol. Melting-point is 47.5°C. (117.5°F.).

Hydrocinnamic acid is employed in the treatment of phthisis; the dose being 10 drops of the alcoholic solution (1 to 5).

PHENYL-SALICYLATE. See Salol.

PHENYL-SALICYLIC ACID. $C_6H_3(OH)(C_6H_5)COOH$.

Synonym: Ortho-oxydiphenyl-carboxylic acid.

This occurs as a white powder, only slightly soluble in water, more soluble in alcohol and glycerin. Phenyl-salicylic acid is employed as an antiseptic dusting powder. •

PHENYL-URETHANE. See Euphorin.

PHILOPAIDIA.

A proprietary diphtheria remedy.

PHOENIXIN. C Cl₄.

Another name for Carbon-Tetrachloride, a non-inflammable liquid solvent.

PHOSPHERGOT.

Generic name given to a mixture of sodium phosphate and ergot, recommended in general debility. It appears in the following three modifications: The variety intended to be taken as a mixture (in sweetened water) contains 1.5 gm. (23 grains) of sodium phosphate and 1 gm (15 grains) of powdered ergot; this quantity constituting a daily dose. In the case of phosphergot powder, each dose represents 0.25 Gm. (3¾ grains) each of dried sodium phosphate and powdered ergot; and this dose is intended to be taken in the morning on an empty stomach. For pills, the following proportions are used: Dried sodium phosphate and extract of ergot, of each 2 Gm. (31 grains); make into 20 pills. Two to four pills to be taken daily.

PHOSPHERRIN.

A mixture of ferric chloride, phosphoric acid and glycerin.

PHOTOXYLIN.

This is a nitrocellulose prepared by the action of nitric acid on wood-wool. When dissolved in a mixture of ether and alcohol (equal parts) a preparation very similar to collodion is obtained. A 3 to 5% solution of photoxylin forms a thick liquid, which leaves on evaporation a much stronger film than collodion.

PICROL. $C_6HI_2(OH)_2SO_3K$.

Synonym: Di-iodo-resorcin-monosulphonate of potassium.

Obtained by the action of a solution of hydriodic and iodic acids on resorcin-monosulphonate of potassium. It forms a colorless, inodorous, very bitter, crystalline powder, which is soluble in water, glycerin, ether and collodion. Contains 52 per cent. of iodine, and is offered as a substitute for iodoform, its antiseptic powers being equivalent to those of corrosive sublimate.

PICROPODOPHYLLIN.

A crystalline principle obtained from the rhizome of the *Podophyllum peltatum* It is claimed to be the active principle of podophyllotoxin.

PINAPIN.

A fermented pine-apple juice. Recommended in treatment of catarrh of the stomach, also in nasal catarrh; in the latter case pinapin is applied as a spray.

PINOL.

The volatile oil prepared from the needles of the *Pinus pumilio*.

PIPERAZIDINE. See Piperazine.

PIPERAZINE. NH-(CH₂)₄-NH.

Synonyms: Diethylene-diamine; Æthylen-imine; Piperazidine; Disper-mine (Spermine).

By the interaction between æthylene bromide and aniline in the presence of a solution of potassium hydrate, di-phenyl-piperazine is formed; this is converted into piperazine by subsequent treatment with nitric or sulphuric acid and distillation of the resulting product with alkalies. Piperazine forms white, deliquescent scales.

which melt at 104° to 107° C. (219.2° to 224.6° F.), boils at 145° C. (293° F.) and is very soluble in water. Because of its property of uniting with uric acid and forming a soluble compound, it is employed as a remedy in treatment of uræmia, rheumatism and gout. Dose 1 Gm. (15 grains) dissolved in about 1 pint of carbonated water. Sometimes given together with phenocoll, when about 1 Gm. of each is dissolved in a pint of carbonated water, and the two solutions mixed.

PIPERINE. $C_{17}H_{19}NO_3$.

An alkaloid which occurs in the fruits of *Piper nigrum, Schinus Mollis,* etc. Obtained synthetically by heating solutions of piperidin and piperic acid in benzol. Forms colorless, almost tasteless, monoclinic prisms, which melt at 128° to 129° C. (262.4° to 264.2° F.), slightly soluble in water and soluble in 30 parts of cold alcohol. Piperine which is contaminated with resin possesses a more or less pungent taste. Employed as antiperiodic and antipyretic in doses of 0.06 to 0.64 Gm. (1 to 10 grains).

PIPERONAL (OR HELIOTROPIN.) $C_8H_6O_3$.

Piperine is converted into potassium piperate by boiling with its equal weight of potassium hydrate and 5 parts of alcohol in a flask with inverted condenser, on cooling, the crystalline mass that separates is washed with alcohol and crystallized from boiling water. One part of potassium piperate is dissolved in 50 parts of hot water, adding slowly in portions an aqueous solution of 2 parts of potassium permanganate, shaking constantly; a soft mass separates, which is strained off and washed with water until free from heliotrope odor. The mixed liquids are distilled, collecting the first distillate separately, because of the presence of the greater portion of the piperonal, which crystallizes out on exposure to cold, while from the latter and weaker distillate it is removed by agitating with ether. Piperonal forms small white crystals, soluble in alcohol and ether, insoluble in cold water. Employed as an antiseptic and antipyretic in doses of 0.5 to 1 Gm. (8 to 15 grains). Because of its heliotrope odor it is employed largely in perfumery.

PIPEROVATINE. $(C_{16}H_{21}NO_2.)$

A crystalline substance obtained from *Piperovatum.* It is insoluble in water, dilute acids and alkalies, soluble in alcohol. Piperovatine acts as a temporary depressant of both motor and sensory nerves, and also as a heart poison. It produces a powerful stimulant effect on the spinal cord, causing a tonic spasm resembling that of strychnine. It, therefore, seems likely to be of service in therapeutics.

PIXOL. See under Cresol.

PODOPHYLLOTOXIN.

An amorphous principle obtained from podophyllin (resin) by extraction with chloroform and precipitation with ether. Forms a white amorphous powder soluble in hot water, dilute alcohol, chloroform and ether. Podophyllotoxin is the active constituent of the mandrake root and its rosin (podophyllin), the latter containing 20 to 30 per cent. of podophyllotoxin. This principle is very active and should be employed with caution; dose 0.001 to 0.01 Gm. (1-64 to 1-6 grain).

POLYSOLVE. See Sodium Sulpho-ricinate.

POTASSIUM SALTS.

Only the most important of the newer preparations are enumerated.

AURO-CYANIDE. $KAuCn_4$. Forms colorless, very soluble crystals, which are employed subcutaneously.

CANTHARIDATE. $C_{10}H_{14}K_2O_6+2H_2O$. Obtained by neutralizing cantharidin with potassium hydrate and evaporating to dryness. It forms a white, very soluble crystalline mass, recommended by Liebreich in form of subcutaneous injections in treatment of tuberculosis (v. Cantharidin).

COBALTO-NITRITE. $K_6CO_2(NO_2)_{12}+2H_2O$. Minute yellow crystals which are slightly soluble in water, insoluble in alcohol. Employed in dyspepsia, cardiac albuminuria, etc. Dose is 0.032 Gm. (½ grain).

DITHIOCARBONATE. K_2COS_2, is prepared by interaction between boiling solutions of carbon disulphide and potassium hydrate. It forms a reddish, deliquescent salt, soluble in alcohol and water. Employed in skin diseases as 5 per cent, ointment or solution.

IODATE. See Iodic Acid.

OSMATE forms a red, soluble, crystalline powder. It is used in combination with bromides in treatment of epilepsy, subcutaneously for neuralgia and goitre. Dose, 0.001 Gm. (1-64 grain).

SOZOIODOL. $C_6H_2I_2(OH)SO_3K+2H_2O$. Forms colorless prisms, which are soluble in 84 parts of water and 200 parts of alcohol. Employed in skin diseases, either as a dusting-powder (3%) diluted with talcum, or in ointments (3%).

SULPHOPHENATE (Sulpho carbolate) $C_6H_5KSO_4+H_2O$, occurs as white shining crystals, readily soluble in water. It possesses antiseptic properties, analogous to sodium sulphophenate.

TELLURATE. K_2TeO_4. A white **crystalline salt, soluble in water.** Employed in phthisis for arresting night sweats.

POWDER, SERUM. See Serum Powder.

PROPYLAMINE. CH_3-CH_2-CH_2-NH_2.

This base should not be confused with trimethylamine $(CH_3)_3N$. Propylamine occurs in colorless crystals of strong ammoniacal odor, melting at 50° C. Recommended in doses of 2 to 4 Gm. per day in treatment of chorea. Best administered with a syrup of peppermint as corrigent.

PROSTADEN.

An extract of the prostate glands, administered in treatment of diseases of these glands.

PSEUDOHYOSCYAMINE. $C_{17}H_{23}NO_3$.

A new alkaloid recently discovered in the *Duboisia myoporoides*. It occurs in yellow needles which melt at 133° to 134° C. (271.4° to 273° F.), sparingly soluble in water. Pseudohyoscyamine is employed in treatment of mania and hysteria, in subcutaneous doses of 0.0005 to 0.006 Gm. (1-120 to 1-10 grain).

PUNICINE. See Pelletierine.

PYOKTANIN (blue). $C_{24}H_{28}N_3Cl$ and $C_{26}H_{30}N_3Cl$.

Synonyms: Pyoktaninum cæruleum; Methyl-violet.

The methyl violet of commerce (a dye stuff) is a mixture of the hydrochlorides of penta and hexa-methyl-para-rosaniline, which forms a very soluble, blue, crystalline powder. Employed as an antiseptic in surgery in the same dilutions as given under yellow pyoktanin (q. v.) It has been employed with success in the treatment of diphtheria, in which the membranes have been painted with a warm saturated solution. Incompatible with corrosive-sublimate; **also** alkalies.

PYOKTANIN (yellow). $C_{17}H_{24}N_3OCl.$

Synonyms: Pyoktaninum aureum; **Auramine**; Imido-tetramethyl-di-p-amido-diphenyl-methane.

Yellow pyoktanin, or auramine, is a yellow dye obtained by the interaction between tetramethyl-diamido-benzophenone, ammonium chloride and zinc chloride at 150° C. It forms a bright yellow powder which is almost insoluble in cold, but very soluble in hot water, also soluble in alcohol. Auramine is employed as an antiseptic for ophthalmic purposes, and in surgery in aqueous solutions (1 to 4 : 10,000), as an antiseptic dusting-powder (1 to 2%) and ointment (2 to 10%). Its aqueous solutions **are** decomposed when heated over 70° C.

PYRAZOL. (Phenyl-methyl-pyrazol-carbonic acid.)

A crystalline compound of composition similar to antipyrine, but used as a diuretic in doses of 1 to 2 Gm. (15 to 30 grains).

PYRAZOLIN. See Antipyrin.

PYRETINE.

Found by Walter to consist of a mixture of acetanilid 58.7 parts, caffeine 6.74 parts, sodium bicarbonate 19.5 parts, and calcium carbonate 13.5 parts. Another analyst finds potassium bromide present. Recommended as an antipyretic.

PYRIDINE. C_5H_5N.

A liquid base which is formed by the dry distillation of nitrogenated organic substances, being found in Dippel's animal oil and coal tar oil. Pure pyridine is a colorless liquid, of peculiar empyreumatic odor, pungent taste, boiling at 117° C. (242.6° F.), miscible with water, with acids forms salts, being the basis of several natural alkaloids. Pyridine is employed in asthma, from 3 to 5 Cc. being poured on a plate and placed in the room with the patient. Internally, in doses of 3 drops, it acts **as a** cardiac stimulant.

PYRODINE. See Hydracetine.

PYROGALLOPYRINE. See under Antipyrine.

PYROZONE.

A name given to represent absolute anhydrous H_2O_2, the active part of that which was formerly called Peroxide of Hydrogen; hence the 3% portion of Aqua Hydrogenii Dioxidi, U. S. P.

The Greek word "pyr" and the word "ozone" were combined to describe the action resulting from the contact of well diffused animal organic matter, like silk or camel's' hair with the thick syrupy fluid of pure H_2O_2. After such articles are moistened with this active princip'e and then slightly warmed on a steam bath, they ignite and burn furiously, as substances do in oxygen, presumably producing both fire and ozone; therefore true H_2O_2 was called Pyrozone.

PYROZONE 3% SOLUTION is an aqueous solution of H_2O_2 of correct pharmacopœial strength. The manufacturers claim that it is more stable than ordinary products of the U. S. P. process, that it is free from barium and other earthy salts and contains only 1-20 of 1% of acidity.

PYROZONE 5% SOLUTION ETHEREAL, is a surgical antiseptic and cleansing bleacher for tooth cavities prior to filling. For external use only.

PYROZONE 25% SOLUTION ETHEREAL, is employed in surgery, dermatology and dentistry as an antiseptic and caustic in a new sense of that term; it destroys pathological cells and apparently does not destroy physiological cells, and has been found by dentists to be a prompt bleacher for teeth discolored with organic matter. For external use only.

QUEBRACHIN. $C_{21}H_{26}N_2O_3$.

One of six alkaloids found in the bark of *Aspidosperma quebracho*. Occurs in pale yellow needles, insoluble in water and but slightly soluble in alcohol and ether. With acids it forms crystalline soluble salts, the *hydrochloride* being most frequently employed. Used internally and subcutaneously in dyspnœa; dose 0.05 to 0.1 Gm. (3-5 to 1½ grains).

QUININE ARSENITE. $C_{20}H_{24}N_2O_2.H_3AsO_3+2H_2O$.

Obtained by dissolving 34 parts of freshly precipitated quinine in a hot aqueous solution of 10 parts of arsenous acid, evaporating at 20° C. (68° F.) and crystallizing. This is a white crystalline powder, containing 69 per cent. of quinine (theoretically), slightly soluble in cold and very soluble in hot water. Employed as antiperiodic in doses 0.005 to 0.03 Gm, (1-12 to ½ grain).

QUININE DI-HYDROCHLORIDE CARBAMATE. $(C_{20}H_{24}N_2O_2).HCl+CO(NH_2)_2HCl.$

Synonyms: Chininum amidato-bichloratum; Chininum bimuriaticum carbamidatum.

This double salt is obtained by dissolving molecular quantities of quinine hydrochloride and urea chloride ($CO(NH_2)_2.HCl$) in boiling water and crystallizing. Forms colorless, soluble, crystals, containing 70 per cent. of quinine. Especially adapted for subcutaneous injections.

QUININE FERRI-CHLORIDE.

This forms dark-brown scales or reddish-brown, hygroscopic powder soluble in water and alcohol. Recommended as hæmostatic for external and internal use; externally dusted over the bleeding surface; used in a 2% solution in uterine hæmorrhage.

QUININE HYDRO-CHLOR-SULPHATE. $(C_{20}H_{24}N_2O_2)_2HCl.H_2SO_4+3H_2O$.

Quinine hydrochloride and bisulphate are dissolved, in molecular proportions, in warm water, evaporated and crystallized. Colorless crystals, soluble in 1 part of water; recommended for subcutaneous injection.

QUININE HYDRIODATE. See Iodic Acid.

QUININE SALICYLATE. $C_{20}H_{24}N_2O_2.C_7H_6O_3+H_2O$.

An alcoholic solution of freshly precipitated quinine is saturated with an alcoholic solution of salicylic acid and crystallized, or 10 parts of quinine sulphate and 3½ parts of sodium salicylate are added to 120 parts of water and heated to boiling; on cooling the quinine salicylate crystallizes out. Forms fine, white needles, soluble in 225 parts of water. Employed as an antipyretic in typhus; also used in rheumatism and gout in doses of 0.1 to 0.5 Gm. (1½ to 8 grains)

QUININE TANNATE.

Obtained by precipitating an aqueous solution of a quinine salt with tannic acid. Forms a yellowish-white amorphous powder, of only a slightly bitter and astringent taste, containing 30 to 32% of quinine, and only very slightly soluble in water. Employed in diarrhœa, whooping-cough, etc., in doses of 0.2 to 0.5 Gm. (3 to 8 grains).

QUINOSOL.

A quinoline compound said to possess bactericidal and antiseptic properties of considerable power. It is reported to act as an antiseptic in solutions as dilute as 1 to 40,000.

QUIONIN.

A substitute offered for quinine. It is said to consist of 90 per cent. of side "bark" alkaloids and a small percentage of quinine.

RESACETIN.

An unknown salt of oxy-phenyl-acetic acid.

β-RESALGIN (Resorcylalgin).

This is obtained by the action of potassium resorcylate on antipyrin, the former being obtained by heating resorcin 1 part and potassium bicarbonate 5 parts, together with 10 parts of water. Resalgin forms crystalline needles, which melt at 115° C. (239° F.), soluble in 150 parts of cold and 20 parts of boiling water, and readily soluble in alcohol, ether and chloroform. Nothing definite is known concerning its medicinal properties and dose.

RESOL. See under Cresol.

New disinfectant obtained by saponifying 1,000 parts of tar with 200 parts of caustic potash, and adding 200 parts of any indifferent substance, methylic alcohol, for instance.

RESOPYRIN. $C_{11}H_{12}N_2O+C_6H_4(OH)_2$ (?).

A solution of 11 parts of resorcin in 30 parts of water is mixed with a solution of 30 parts of antipyrine in 90 parts of water, the resulting precipitate (resopyrin) is collected and recrystallized from alcohol. Forms colorless crystals, insoluble in water and soluble in alcohol. The medicinal properties of this body have not been determined.

RESORBIN.

This is proposed as an ointment base, being prepared by emulsifying almond oil and water with yellow wax, gelatine and soap; it is said to possess great penetrating power, but is open to the objection that ointments prepared with it do not keep.

RESORCIN. $C_6H_4(OH)_2$.

Synonyms: Resorcinol; Meta-dihydroxy-benzene.

This is obtained on fusing many resins (umbelliferous), also m-phenol-sulphonic acid, also m-benzene-disulphonic acid, with potash; the fused mass is acidified with hydrochloric acid and the resorcin extracted with ether. For description see U.S.P. Internally it is employed in gastritis, gastric ulcers, affections of the larynx generally; dose, 0.2 to 0.5 Gm. (3 to 8 grains). Externally in diphtheria as a resorcin glycerole (10%) for topical application, as an ointment (5 to 25%) in various skin diseases.

RESORCINOL.

This term "resorcinol" has been applied by chemists to resorcin, in compliance with the rules of modern chemical nomenclature. Unfortunately, the same title has been given to an amorphous, brown powder which is prepared by triturating together equal parts of resorcin and iodoform. It is recommended as an application to gangrenous sores, ulcers, chancres, etc.; diluted with starch it is used as a dusting powder (5%); also as ointment of 5 to 15% strength. It is soluble in ether and partly soluble in alcohol and water. The caustic action of the resorcin and the unpleasant odor and toxic effects of the iodoform are lost in this preparation. Resorcinol is recommended as an antiseptic.

RETINOL. (CODOL.)

Obtained as a product of the destructive distillation of resin (colophony). Retinol forms a yellowish, fluorescent, oily liquid of sp. gr. 0.900, boiling between 240° to 280° C. (464° to 536° F.). Employed as a solvent for various organic bodies as iodol, aristol, camphor, cocaine, codeine, phenol, phosphorus, creosote, etc.

RHEIN. See Chrysophanic Acid.

RHINALGIN.

A nasal suppository containing cacao butter, alumnol, menthol and oil of valerian. Recommended in coryza.

RHODALLIN. See Thiosinamine.

RUBIDIUM-AMMONIUM BROMIDE. $(RbBr.3NH_4Br.)$

This double salt forms a yellowish-white powder of a saline taste, soluble in water, used as a sedative and hypnotic in epilepsy in doses of 4 to 6 Gm. (60 to 90 grains).

RUBIDIUM IODIDE. Rb I.

Forms colorless cubical crystals which are readily soluble in water. Employed in the same instances where the potassium or sodium iodide is indicated, possessing the advantage of not causing gastric and other disturbances which usually accompany the administration of these salts in large quantities. Dose, 0.13 Gm. (2 grains).

RUBIDIUM TARTRATE ($RbC_4H_5O_6$) and BROMIDE.

Form transparent soluble crystals, and, like the corresponding cæsium salts, these are serviceable in cardiac palpitations of nervous origin. The dose of either is 0.18 to 0.30 Gm. (3 to 5 grains).

RUBROL.

A solution of boric acid, thymol, and a coal tar derivative (?) in water, recommended as a gonorrhœal injection.

RUMICIN.   See Chrysophanic Acid.

SACHARIN. $C_6H_4 < {CO \atop SO_2} > NH$.

Synonyms :   Benzoyl-sulphonic-imide ;   Orthosulphamine-benzoic-anhydride ; Gluside ; Glucusimide.

Saccharin is an intensely sweet principle prepared from toluene ($C_6H_5CH_3$) by first converting this into the mixture of mono-sulphonic acids, which, by the action of phosphorus pentachloride, are converted into the corresponding toluene-sulphonic chlorides. By the action of ammonia the ortho compound is converted into sulphamin benzoic acid, which by oxidation yields the above imide (Saccharin). The pure ortho compound forms a white crystalline powder, which possesses 500 times the sweetening power of cane sugar ; it is soluble in about 400 parts of water (15° C.), more so in alcohol and glycerin (1:30), readily soluble in water in presence of alkalies ($NaHCO_3$). Mixed with water and neutralized with sodium bicarbonate, it forms the soluble sodium salt, " soluble gluside" or "soluble saccharin." The chief use of saccharin is as a sweetening agent in the food of diabetic patients. A syrup of saccharin may be prepared by dissolving saccharin 10 Gm., sodium bicarbonate 12 Gm., in 1,000 Cc. of water. Syrup of saccharin may be employed in many mixtures where cane-sugar syrup is inadmissable.

SALACETOL. $CH < {OH \atop COOCH_2COCH_3}$.

Synonym : Salicyl-Acetol.

A compound differing from salol in the replacement of the phenyl group (C_6H_5) by the acetone radical (CH_3-CO-CH_3); introduced as a substitute for salol in order to avoid the elimination of phenol in the organism. It is prepared by interaction between monochlor-acetone and sodium salicylate. Forms fine needle-like crystals or scales, melting at 71° C. (159.8° F.), insoluble in cold water, slightly soluble in cold alcohol, freely soluble in hot alcohol, ether and chloroform. By action of alkalies it yields up its salicylic acid (about 71%). Salacetol is employed in all instances where salol is indicated, in doses of 2 to 3 Gm. (30 to 45 grains).

SALACTOL.

A preparation consisting of the sodium salts of salicylic and lactic acids has been introduced under this name, and when dissolved in a 1% solution of hydrogen peroxide it is recommended as an efficient remedy for diphtheria. The solution is applied to the throat with a brush every four hours, and in the intervals the solution is used as a gargle. It is also stated to act as a prophylactic.

SALAZOLON.   Same as Salipyrin.   See under Antipyrine.

SALICYL-ACETOL.   See Salacetol.

SALICYLAMIDE. $C_6H_4 < {OH \atop CONH_2}$.

This compound differs from salicylic acid in the replacement of the hydroxyl of the carboxyl group by the amido radicle (NH_2). Obtained by the action of concentrated ammonia on salicylic-methyl-ester (oil of wintergreen), yielding colorless, inodorous and tasteless crystals, melting at 138° C. (280.4° F.), soluble in 250 parts of water, readily so in alcohol and ether. Salicylamide possesses the same therapeutical properties as salicylic acid, having the advantage of being tasteless, more soluble and acting more readily in smaller doses. Dose, 0.13 to 0.3 Gm. (2 to 5 grains).

SALICYL-ANILID.   See Salifebrin.

SALICYL-PHENETIDINE. See Malakin.
SALICYL-*p*-PHENETIDINE. See Saliphen.
SALICYL-SULPHURIC ACID. See Sulpho-Salicylic Acid.
SALICYLIC-ALCOHOL. See Saligenin.
SALICYLIC NAPHTHYLIC ETHER. See Betol.
SALICYLIDEN PHENETIDINE. See Malakin.
SALIFEBRIN, OR SALICYLANILID.

A preparation of salicylic acid and acetanilid, in which both constituents are probably fused together and powdered. Forms a white powder, soluble in alcohol, insoluble in water. Recommended as an antineuralgic and antipyretic. No authoritative dose has been given.

SALIGENIN. (Salicylic alcohol.) $C_6H_4(OH)CH_2OH$.

An oxy-benzoyl-alcohol, obtained by the action of acids or ferments (emulsin, saliva, etc.) on salicin, a glucoside. This same reaction takes place in the human organism when salicin is taken internally, yielding, however, only 43 per cent. of saligenin. A dose of 12 Gm. (3 drachms) of salicin corresponds to about 4.2 Gm. (60 1-3 grains) of saligenin. It will be seen that saligenin, which constitutes the activity, is an excellent substitute for salicin in the treatment of malaria, rheumatism, typhus, etc. Saligenin crystallizes in colorless scales or needles, melting at 86° C. (186.8° F.), of a slightly bitter taste, soluble in alcohol and water. It is now prepared synthetically through the condensation of phenol with formaldehyde.

SALINAPHTHOL. See Betol.

SALIPHEN. $C_6H_4 < \frac{OC_2H_5}{NH.C_6H_4(OH)CO}$.

Synonyms: Salicyl-p-phenetidine.

Obtained by the action of salicylic acid on phenetidine in the presence of phosphorus trichloride. Forms colorless crystals, melting at 139.5° C. (283° F.), insoluble in water, but soluble in alcohol. Its slight antifebrile action has not brought it into any favor.

SALIPYRAZOLIN. Same as Salipyrin. See under Antipyrin.

SALITHYMOL. $C_6H_4 < \frac{OH}{COO.C_{10}H_{13}O}$.

A thymol ester of salicylic acid, prepared by the action of phosphorus trichloride on molecular quantities of sodium salicylate and thymol sodium. Salithymol forms a white crystalline powder, of sweet taste, insoluble in water, very soluble in alcohol and ether. It is recommended as an antiseptic.

SALOCOLL. See under Phenocoll.

SALOL. $C_6H_4 < \frac{OH}{COO C_6H_5}$.

Synonym: Phenyl Salicylate.

This phenyl ester of salicylic acid is obtained by the action of sodium salicylate on sodium phenylate in the presence of phosphorus oxychloride or phosgene. The reaction product is thoroughly washed with water and crystallized from alcohol. For description see U. S. P. Salol, when taken, passes unabsorbed through the stomach into the intestines, where, under the influence of alkaline secretions, it is split up into salicylic acid and phenol; to this dissociation its value as an intestinal disinfectant in case of dysentery, cholera, etc., is due. In view of this peculiarity, salol is employed for coating pills* which are intended to act only in the intestines. As an antirheumatic its dose is 1 to 2 Gm. (15 to 30 grains), in diarrhœa and intestinal troubles of children 0.13 to 0.19 Gm. (2 to 3 grains); as an antiseptic and deodorant externally in the form of dusting-powder (1:3), being diluted with starch or talcum, ointment or collodion (1: to ether 4 and collodion 30).

SALOL CAMPHOR is made by mixing 3 parts of salol with 2 parts of camphor, both in fine powder, fusing and filtering the product. It forms a colorless, oily liquid, insoluble in water, soluble in ether, chloroform and the oils; by the action of light and air it undergoes decomposition. Employed locally as an antiseptic.

* COBLENTZ' Handbook of Pharmacy, p. 329.

DI-IODO-SALOL, $C_6H_2I_2(OH)CO_2C_6H_5$. the phenyl ester of di-iodo-salicylic acid, is obtained by the condensation of di-iodo-salicylic acid with phenol. Forms an inodorous, tasteless, crystalline powder, melting at 133° C. (271.4° F.); employed in treatment of skin diseases.

NITRO-SALOL, $C_6H_4(OH)CO_2.C_6H_4NO_2$. salicylic-p-nitro phenyl-ester, is obtained by condensation of salicylic acid with p-nitrophenol. Forms a yellowish, inodorous and tasteless crystalline powder, melting at 148° C. (298.4° F.), insoluble in water, and soluble in alcohol and ether. In the intestines it is split up into its constituents. Employed in the manufacture of salophen.

TRIBROM SALOL $(C_6H_4(OH)COO.C_6H_2Br_2)$ is a valuable intestinal **antiseptic**, soluble in alkaline secretions, that is, it is decomposed into tribromphenol and salicylic acid in passing through the body.

SALOPHEN. $C_6H_4(OH)CO_2.C_6H_4NH.COCH_3$.

Synonyms: Acetyl-p-amido-phenyl salicylate; Acet-p-amido-salol.

This body was introduced as a substitute for salol in order to avoid effects resulting from the liberation of phenol in the organism, salophen being split up in the intestines into salicylic acid and acetyl p-amidophenol. Obtained by the reduction and acetification of salicylic-p-nitro-phenol (nitro salol). $C_6H_4(OH)CO_2.C_6H_4NO_2$. yielding colorless crystals melting at 187° to 188° C. (368.6° to 370.4° F.), insoluble in water, soluble in alcohol and ether, the alcoholic solution being colored violet by ferric chloride. Employed as an antineuralgic and antirheumatic in doses of 1 to 2 Gm. (15 to 30 grains). Lately recommended as a specific in influenza with nervous complications, in doses of 0.5 Gm. (8 grains) every two hours. For children, doses of 0.25 Gm. (4 grains), **every three hours.**

SALUBRINE.

A composition hailing from Sweden, and containing, according to Hager, 2 per cent of anhydrous acetic acid, 25 per cent. of acetic ether, 50 per cent. of alcohol, and the balance of distilled water. It is antiseptic, astringent and hæmostatic, and is used, diluted with water, as a gargle, and on compresses.

SALUMIN (insoluble). $(C_6H_4(OH)COO)_6Al_2+3H_2O$.

Synonym: Aluminum salicylate.

Obtained as an insoluble precipitate by interaction between solutions of a salt of aluminum and sodium salicylate. Forms a reddish-white, insoluble powder, which is employed as a dusting-powder in catarrhal affections of the nose and pharynx.

SALUMIN (soluble). $(C_6H_4(ONH_4)COO)_6Al_2+2H_2O$.

Synonym: Aluminum-ammonium salicylate.

On treating "salumin" with ammonia "soluble salumin" is obtained. Employed for same purposes as Salumin.

SANATOL. See under Cresol.

SANGUINAL.

Prepared by defibrinating fresh blood and evaporating to a pilular extract, in which condition it is made up into pills, each of which is said to represent 5 Cc. of fresh blood. The composition of the extract is said to be 46 parts of peptonized muscle albumin, 44 parts of blood salts and 10 parts of oxyhæmoglobin.

SANTONINOXIM. $C_{15}H_{18}O_2.NOH$.

Santonin 5p., hydroxylamine hydrochlorate 4p., calcium carbonate 4p., and alcohol 50p., are boiled from 6 to 7 hours in a flask with reflux condenser,* filtering and pouring into 5 times its volume of water, whereby santoninoxim separates. Forms white crystals, insoluble in cold water, soluble in alcohol, ether, fats and fatty oils, melts at 216° to 217° C. (420.8° to 422.6° F). Because of its comparatively non-toxic nature it is preferred to santonin as an anthelmintic; dose for children from 2 to 3 years 0.06 Gm. (1 grain), from 4 to 6 years 0.09 Gm. (1½ grains), from 6 to 9 years 0.13 Gm. (2 grains) in two doses about one hour apart, followed by a purgative.

SAPOCARBOL. See under Cresol.

SAPROL. See under Cresol.

---

* Coblentz' Handbook of Pharmacy, p. 94.

SCLEROTIC ACID (Dragendorff's).

A faintly acid hygroscopic powder obtained from ergot, it is inodorous, tasteless and readily soluble in water. Recommended for injection as a substitute for ergotin in epilepsy, inferior to ergot in gynæcology. Dose, ½ grain.

SCOPOLAMINE. See Hyoscine.

SECALOSE.

A carbohydrate obtained from green rye. It is soluble in water, from which it is precipitated by alcohol. Desiccated over sulphuric acid it forms a white powder which is very hygrometric. By inversion it is converted into lævulose.

SEDATIN. See under Phenetidine.

SENECINE.

An elixir prepared from the *Senecio Jacobæa*, recommended as an emmenagogue, Not to be confused with the alkaloid of the *Senecio vulgaris*.

SEPTENTRIONALIN.

An alkaloid prepared from the *Aconitum septentrionale*, recommended as an antidote in strychnia poisoning, also in treatment of tetanus and hydrophobia.

SERUM PASTE.

The freshly prepared serum from ox-blood is thoroughly mixed with 25 per cent. of zinc oxide and sterilized at 70° C. in a thermostat. When painted over denuded or diseased surfaces, it dries readily, leaving a flim which may be readily removed by washing with water.

SERUM POWDER.

A mixture of freshly prepared serum and zinc oxide (25%) is spread on glass plates and dried, then finely powdered and sterilized at 100° C. Recommended as an antiseptic dusting-powder to be employed alone or mixed with iodoform.

SILVER FLUORIDE. (AgFl.)

This forms a brown, glassy hygroscopic mass, very soluble in water. Used as an antiseptic in anthrax infections.

SILVER IODATES. See Iodic Acid.

SODIUM SALTS.

Only the most important of the new salts are enumerated.

ÆTHYLATE, $CH_3.CH_2ONa$, is formed by the action of metallic sodium upon absolute alcohol. It forms a white powder, of caustic taste, soluble in alcohol and water. Employed in treatment of skin diseases, as psoriasis, lupus, etc., painting the parts with a 10 per cent. aqueous solution.

ANISATE, $C_6H_4(OCH_3)COONa$. Anisic acid is obtained by oxidizing anethol (main constituent of anise oil) with a mixture of sulphuric acid and potassium bichromate; the sodium salt is made by neutralizing an aqueous solution of this acid with sodium carbonate, evaporating and crystallizing. Sodium anisate forms a soluble crystalline powder, which is recommended as a substitute for sodium salicylate, being an antirheumatic and antipyretic.

AURO-CHLORIDE, $AuCl_3.NaCl + 2H_2O$. A double salt of gold and sodium chloride, forming a golden-yellow hygroscopic powder, readily soluble in water and partly in alcohol. Employed in syphilitic diseases, the dose being 0.016 to 0.06 Gm. (¼ to 1 grain).

CHLOROBORATE is obtained by reaction between boron terchloride and sodium hydrate. It forms a soluble white crystalline powder, possessing powerful antiseptic properties.

CINNAMATE ($C_6H_5CH=CHCOONa$), a white crystalline powder, soluble in water, Recommended in 5 per cent. sterilized solutions, hypodermically and internally in treatment of tuberculosis.

DI-IODO-SALICYLATE, $[C_6H_2(OH)I_2.+COONa]_2 + 5H_2O$. Diiodosalicylic acid is obtained by the action of iodine and iodic acid on salicylic acid in alcoholic solution; the sodium salt of this acid is obtained by neutralization with sodium carbonate. This salt forms white crystalline scales, which are soluble in 50 parts of water. Employed as an analgesic, antipyretic and antiseptic in doses of 0.5 to 1 Gm. (8 to 15 grains).

DI-THIO-SALICYLATE, I. and II. See Di-thio-salicylic Acid.

ICHTHYOL-SULPHONATE. See under Ichthyol.

IODATE. See Iodic Acid.

NAPHTHOLATE. See Microcidin.

PARA-CRESOTATE. This is a fine, white, crystalline powder, of a bitter taste, soluble in 24 parts of water. It is employed as an antiseptic and antirheumatic, the daily total dose being 2 to 6 Gm. (30 to 90 grains).

PHENOL-SULPHO-RICINATE, is a solution of 4 parts of sodium ricinate (q. v.) in 1 part of carbolic acid. A caustic fluid recommended in treatment of diphtheria.

PHENOSUCCINATE. The sodium salt of phenosuccin is prepared by warming succinamin with soda solution. It forms a white powder, readily soluble in water. It is to be preferred to phenosuccin itself from a therapeutic point of view, and may be administered in doses of 0.5 to 3 Gm. (7½ to 46 grains) as an antipyretic and antineuralgic.

SILICO-FLUORIDE. (NaF)$_2$SiF$_4$. A white crystalline powder, which is only very slightly soluble in water. Employed in aqueous solution (2 : 1,000) as an antiseptic wash.

SULFOCAFFEATE. Since the introduction of the sulpho group decreases the medicinal potency in phenol groups, the same was tried here with success. Bitter, crystalline, slightly soluble in cold water; non-toxic, does not irritate the stomach. Solutions containing more than 5% are not stable. Besides above sodo, lithium and strontium salts are prepared. A powerful diuretic. Dose, 1 Gm. in capsule.

SULPHO-RICINATE, (Solvin, Polysolve). By the action of concentrated sulphuric acid on the triglycerides of the fatty acids, or the fatty acids themselves, sulphoricinic acid is formed; this on neutralization with sodium hydrate gives the above named salt. This is a brownish, syrupy liquid, which is soluble in alcohol and water. Employed as a solvent for iodine, iodoform, etc.

SULPHOSALICYLATE. C$_6$H$_3$(OH)$<^{COOH}_{SO_3Na}$. Salicyl sulphonic acid is obtained by the action of sulphuric acid on salicylic acid ; this product is then only partly neutralized with sodium carbonate, resulting in the saturation of the sulphonic acid group only. This salt forms a white crystalline powder, of a slightly acid and astringent taste, soluble in 25 p. of water and insoluble in alcohol and ether. Proposed as a substitute for sodium salicylate.

SULPHO-THIOPHENE. See Thiophene Sulphonate.

SULPHOTUMENOLATE. See Tumenol.

TELLURATE. See Potassium Tellurate.

TETRABORATE. This salt contains 56% each of boric acid and sodium biborate. It is readily soluble in water and is used as an antiseptic in place of boric acid, usually applied locally in 16% solutions.

THIOPHENE-SULPHONATE. C$_4$H$_3$S—SO$_3$Na, is obtained by neutralizing thiophene-sulphonic acid with sodium carbonate. It is a white crystalline powder, of unpleasant odor, soluble in water, and contains 33% of sulphur. Employed in prurigo, as a 5 to 10% ointment.

TUMENOL-SULPHONATE. See Tumenol.

SOLANIN. C$_{42}$H$_{87}$NO$_{18}$.

A principle which occurs in the berries, flowering tops and fruits of various solanaceous plants. Obtained from the aqueous acidulated extract of potato-sprouts by making alkaline with ammonia and shaking with ether. Solanin occurs in colorless acicular crystals, melting at 235° C. (455° F.), bitter taste, insoluble in water, and but slightly in alcohol.

Recommended in doses of 0.01 to 0.06 Gm. (1-6 to 1 grain) as an analgesic in neuralgia, also in bronchitis and asthma.

SOLPHINOL.

A mixture of borax, boric acid and alkali sulphites. Used as an antiseptic.

SOLUTOL. See under Cresol.

SOLVEOL. See under Cresol.

SOLVIN. See Sodium Sulpho-ricinate.

SOMATOSE.

A preparation in which the albumenoids and nutritive constituents of flesh are converted into soluble albumoses, 5 parts of somatose representing 30 parts of beef in nutritive value. Forms a pale yellowish powder which is readily soluble in water, forming an almost odorless and tasteless solution.

Employed as a food for patients afflicted with **weak digestion, being given 15 to 30 Gm.** (or ½ to 1 ounce) in milk, cocoa or soup.

SOMNAL. $C_7H_{12}Cl_3O_2N.$

An ethylated compound of chloral and urethane, forming a clear colorless liquid of burning taste. Recommended as a hypnotic in doses of 15 to 30 minims.

SOZAL. $(C_6H_4(OH)SO_3)_3Al.$

Synonym: Aluminum-para-phenol-sulphonate.

Obtained by dissolving aluminum hydroxide in para-phenol-sulphonic acid. Forms a crystalline powder, soluble in water, glycerine and alcohol, possesses an astringent taste and a phenol-like odor. Its aqueous solution is colored violet by ferric chloride, and precipitates albumen, soluble in excess of the latter.

Employed in solution (1%) as a wash for tubercular ulcers and **purulent affections.**

SOZOIODOL. See Sozoiodolic Acid.

SOZOIODOLIC ACID. $C_6H_2I_2(OH)SO_3H.$

Synonyms: Sozoiodol; Di-iodo-para-phenol-sulphonic Acid.

This is obtained by the interaction between a solution of potassium para-phenol-sulphonate in dilute hydrochloric acid and a solution of potassium iodide and iodate (5KI+KIO₃) in molecular proportions. The acid potassium salt which crystallizes out is treated with the necessary amount of sulphuric acid, whereby sozoiodol is liberated.

Sozoiodol crystallizes from water in acicular prisms with 3 molecules of water, readily soluble in water, alcohol and glycerin.

Solutions of this compound give a violet-blue coloration **with ferric salts.** Sozoiodol is employed as an antiseptic, being usually used in 2 or 3 per cent. solutions. It is also employed as a dusting-powder, containing 5 to **10 and 20% diluted with powdered** French chalk or starch.

MERCURY SOZOIODOL, $C_6H_2I_2(OH)SO_3)_2Hg$, forms a lemon yellow powder, obtained by interaction between concentrated aqueous solutions of sozoiodol sodium and mercuric nitrate. This compound is soluble in 500 parts of water, readily soluble in a solution of sodium chloride. It is employed chiefly in the specific treatment of syphilis locally and subcutaneously. The 2.5 per cent. solution, **or 1% ointment or** dusting powder, is the usual strength of **dispensing.**

POTASSIUM AND SODIUM SOZOIODOL, $C_6H_2I_2(OH)SO_3K$ (or Na)+2H₂O, are obtained by saturating the sozoiodolic acid with either potassium or sodium carbonate and crystallizing. Of these the potassium salt is soluble in 50, and the sodium salt in 14 parts of water. The aqueous solutions of these compounds gradually darken on exposure to light. These compounds are employed in like manner to sozoiodolic acid.

ZINC SOZOIODOL, $(C_6H_2I_2(OH)SO_3)_2Zn+6H_2O$, forms colorless crystals, soluble in 20 parts of water; employed in medicine as the above.

SOZOLIC ACID. See Aseptol.

SPARTEINE. $C_{15}H_{26}N_2.$

An alkaloid which occurs with scoparine in the tops of *Spartium scoparium.* It forms a volatile oily liquid which boils at 288° C. (550.4° F.), unites with acids, forming stable crystalline salts.

SULPHATE. $((C_{15}H_{26}N_2)_2H_2SO_4+4H_2O).$ Colorless, odorless, slightly hygroscopic crystals, soluble in water and alcohol.

Employed as a heart-tonic, like digitalis, in **doses of 0.01 to 0.02 Gm. (1-6 to 1-3 grain).**

SPASMOTIN. $C_{20}H_{21}O_9.$

Synonyms: Sphacelotoxin, Spasmotoxine.

A poisonous principle extracted from ergot, soluble in alcohol and ether, insoluble in water. It is said to represent the action of ergot. Dose, 0.032 to 0.1 Gm. (½ to 1½ grains).

SPASMOTOXINE. See Spasmotin.

SPERMIN. (CH₂)₂NH.

A base belonging to the class of leucomaines, obtained from the seminal fluid of animals. A readily soluble, crystalline substance, which is usually obtained in the form of the hydrochlorate. A 2 per cent. solution is employed subcutaneously in quantities of ½ to 1 Cc. (8 to 16 minims) once daily in treatment of nervous diseases complicated with anæmia.

SPHACELOTOXIN. See Spasmotin.

STERESOL.

A brown, thick liquid, obtained by dissolving shellac 270 parts, gum benzoin 10 parts, balsam tolu 10 parts, phenol 100 parts, oil of cinnamon 8 parts, and saccharin 6 parts in alcohol sufficient to make 1,000 parts. Recommended as an antiseptic varnish for tubercular sores and various skin diseases.

STRONTIUM SALTS.

These salts have been recommended as preferable to the corresponding salts of sodium or potassium for the same diseases on the ground of being better borne by the system. The bromide, carbonate, iodide and lactate are most frequently used.

STROPHANTHIN. C₂₀H₃₄O₁₀ or C₃₁H₄₈O₁₂(?

The active principle (a glucoside) of the seeds of the species of *Strophanthus*. Forms a white crystalline powder, melting at 185° C. (365° F.), soluble in 40 parts of water (18° C.), readily in alcohol. Strophanthin is employed as a substitute for digitalis, being free from all disturbing effects upon the respiratory centres and producing less gastric disturbance. Dose, 0.0002 to 0.0006 Gm. (1-300 to 1-200 grain). Very powerful; should be used with great caution. Antidotes, aconite and veratrum viride. Commercial strophanthin is quite variable in strength, hence uncertain in effect.

STYPTICIN.

A yellow crystalline substance, which is especially recommended for checking uterine hemorrhages, 0.2 Gm. (3 grains) of a 10 per cent. solution being injected into the gluteal region. In profuse menstruation 0.035 Gm. (½ grain) of the remedy are administered five times daily for four days before the expected period.

STYRACOL. See under Guaiacol.

SULFINIDUM ABSOLUTUM.

Absolutely pure saccharin, free from isomers. See Saccharin.

SULPHAMINOL. C₆H₄ $<^S_{NH}>$ C₆H₃OH.

Synonym: Thio-oxy-diphenylamine.

Obtained by boiling meta-oxy-diphenylamine with sulphur and caustic soda solution, and precipitating with a solution of ammonium chloride. Sulphaminol forms an inodorous, pale yellow powder, melting at 155° C. (311° F.), insoluble in water, soluble in alkali solutions. alcohol and glacial acetic acid. Employed as a substitute for iodoform, used as a deodorizing antiseptic for both internal and external use; it readily breaks up, yielding phenol and sulphur. Internally employed in cystitis, dose being 0.25 Gm. (about 4 grains).

SULPHONAL. (CH₃)₂C(SO₂C₂H₅)₂.

Synonym: Di-ethyl-sulphon-dimethyl-methane.

Through a mixture of anhydrous ethyl-mercaptan (C₂H₅SH) and acetone (CH₃-CO-CH₃), dry hydrochloric acid gas is passed, resulting in the condensation product mercaptal (di-thio-ethyl-di-methyl-methane), which on oxidation yields sulphonal. Forms colorless permanent crystals, melting at 125° to 126° C. (257° to 258.8° F.), soluble in 500 parts of cold and 15 parts of boiling water, in 65 parts of cold and 2 parts of boiling alcohol.

Employed as a valuable hypnotic in doses of 1 to 2 Gm. (15 to 30 grains).

SULPHO-SALICYLIC ACID. C₆H₃(SO₃H)(OH)COOH.

Synonym: Salicyl-Sulphuric Acid.

This is prepared by the action of fuming sulphuric acid on salicylic acid; it forms white crystals, which are soluble in water and alcohol. Employed as a substitute for sodium salicylate in treatment of articular rheumatism. This compound is a valuable reagent for proteids, albumins and peptones. An albumose or peptone is precipitated, but redissolves on boiling the solution, while albumin or globulin does not.

SULPHO-TUMENOLIC ACID. See Tumenol.

SULPHURIC ACID PASTE.

A caustic application composed of a mixture of equal parts of sulphuric acid and powdered saffron, the latter being employed because of the finely subdivided condition of the carbon yielded.

SYMPHOROL. See Sodium Sulphocaffeate.

SYNDETICON.

A varnish prepared by dissolving 100 parts of fish glue in 125 parts of acetic acid (glacial), mixing with a solution of 20 parts of gelatine in 125 parts of water. This solution is then mixed with 20 parts of shellac varnish (concentrated alcoholic solution of shellac).

TANNAL (Insoluble). $Al_2(OH)_4(C_{14}H_9O_9)_2 + 10H_2O$.

Synonym: Aluminum basic tannate.

Formed by precipitating a solution of an aluminum salt with a solution of tannic acid in presence of an alkali. Tannal is a brownish-yellow, insoluble powder, employed as an astringent in catarrh of the respiratory organs.

TANNAL (Soluble). $Al_2(C_4H_5O_6)_2(C_{14}H_9O_9)_2 + 6H_2O$.

Synonym: Aluminum tannic-tartrate.

Obtained by treating insoluble tannal with tartaric acid, yielding a brownish yellow soluble powder, which is employed for the same purpose as the above.

TANNIGEN.

Synonym: Diacetyl Tannin.

An acetic ester of tannic acid, prepared by the action of acetic anhydride on tannin dissolved in glacial acetic acid. Forms a yellowish-gray, odorless and tasteless, hygroscopic powder, insoluble in water, only slightly soluble in ether, very soluble in alcohol. Its solutions are colored blue-black by ferric chloride and decomposed by alkalies. Tannigen is recommended in treatment of chronic diarrhœa, acting as an intestinal astringent, since owing to its insolubility it passes through the stomach into the intestines, where in presence of the alkaline secretions it is broken up into its constituents.

TARTARLITHINE.

This is an effervescent salt, the lithium analogue of cream of tartar, containing none of the additional alkaline salts common to the granular effervescent preparations. It is recommended as a uric acid solvent, in place of the other salts of lithium, for gout, rheumatism and all the manifestations of uric-acidæmia, and is presented in tablet form.

Dose: One or two of the 5 grain tablets, dissolved in a goblet of water, may be taken on a reasonably empty stomach, four times a day. Tartarlithine is intended to increase the alkalinity of the blood, by giving vegetable acid up to a point where it will contribute an alkalinity more effective than alkalies as such, for the elimination of uric acid.

TARTARLITHINE AND SULPHUR is prepared with equal parts of Tartarlithine and precipitated Sulphur, compressed into 5 grain tablets. Indicated in the treatment of chronic sore throat, chronic bronchitis accompanied with copious secretions; in digestive difficulties due to disordered action of the liver, which ultimately lead to lithæmia and structural lesions, in addition to many benefits as a pulmonary or intestinal disinfectant. This combination of sulphur, probably after absorption, favors the bile-producing function of the liver, since taurocholic acid normally contains a large proportion of sulphur. It is prescribed in diseases of the nails, the scalp, and generally in superficial skin diseases.

Dose: Same as Tartarlithine.

TEREBENE. $C_{10}H_{16}(?)$.

This is produced by the action of concentrated sulphuric acid upon oil of turpentine and repeated distillation for purification. It consists of a mixture of camphene, cymene, borneol and terpilene. For description see U. S. P. Terebene is an agreeable antiseptic, disinfectant and deodorizer, a 5 per cent. aqueous solution forming a very serviceable surgical dressing, while its vapor is inhaled in treatment of bronchial affections and pulmonary tuberculosis. Internally, in doses of 5 to 6 drops in emulsion or tablet form, it acts as an expectorant.

TERPIN HYDRATE. $C_{10}H_{18}(OH)_2 + H_2O$.

A mixture of rectified turpentine oil (4 pts.), alcohol (3 pts.) and nitric acid (1 pt.) is set aside in a shallow porcelain dish for several days; crystals of terpin hydrate separate, and these are recrystallized from 95% alcohol. For description see U.S.P. p. 404.

Employed as expectorant in **bronchitis** and chronic **nephritis**, in doses of 0.2 to 0.4 Gm. (3 to 6 grains).

TERPINOL.

By the distillation of terpin hydrate with dilute sulphuric acid, terpinol is obtained; this consists of a mixture of terpineol ($C_{10}H_{17}OH$) an alcohol, and three terpenes ($C_{10}H_{16}$), terpineue, terpineolene and dipentene. Terpinol is an oily liquid, of hyacinthine odor, boiling at 168° C., sp. gr. 0.852, insoluble in water and soluble in alcohol and ether.

Employed as a bronchial stimulant in doses of 0 5 to 1 Gm. (8 to 15 grains). Terpinol is sometimes used to mask the odor of iodoform.

TERROLINE.

A name for a special brand of petroleum jelly.

TERTIARY AMYL ALCOHOL. See Amylenum Hydratum.

TETRA-ETHYL-AMMONIUM HYDROXIDE. $(C_2H_5)_4N.OH$.

This forms a hygroscopic, crystalline salt, bitter taste, and very soluble in water. Recommended as a uric acid solvent, being administered in doses of 10 to 15 minims of a 10 per cent. solution.

TETRA-HYDRO-BETA-NAPHTHYLAMINE. See Thermin.

TETRA-HYDRO-PARA-CHINANISOL. See Thalline.

TETRA-IODO-PHENOL-PHTALEIN. See Nosophen.

TETRA-IODO-PYRROL. See Iodol.

TETRA-THIO-DICHLOR-SALICYLIC ACID. $(S_2=C_6H(Cl)(OH)COOH)_2$.

This is obtained by heating salicylic acid (37.6 p.) with sulphuryl chloride (55. p.). It forms a reddish-yellow powder, which is soluble in aqueous solutions of the alkalies. It is employed chiefly as an antiseptic dusting powder.

TETRONAL. $(C_2H_5)_2.C.(SO_2C_2H_5)_2$.

Synonym: Di-ethyl-sulphon-di-ethyl-methane.

An analogue of sulphonal and trional, differing in the possession of four ethyl groups, while the former contains two and the latter three. The method of preparation is the same as that of sulphonal, only that di-ethyl-ketone ($C_2H_5-CO-C_2H_5$) is employed in place of acetone. This compound forms colorless, crystalline scales, melting at 89° C. (192.2° F.), soluble in 450 parts of cold water, readily in alcohol and ether.

Tetronal is employed as a hypnotic in doses of 1 to 2 Gm. (15 to 30 grains).

TEUCRIN.

The purified extract of *Teucrium scordium* sterilized in small glass tubes. It forms a dark-brown fluid of pungent taste. Employed in treatment of tuberculous abscesses, fungous adenitis, lupus, etc. producing local active hyperæmia and organic reaction that arrests development of these diseases. Dose, hypodermically, 50 minims ; locally, 10 grains as ointment with lanolin, once daily.

THALLINE. $C_9H_{10}N(OCH_3)$.

Synonym: Tetra-hydro-para-chinanisol.

This liquid base, a chinolin derivative, is obtained by heating a mixture of para-amido-anisol, para-nitro-anisol, glycerin and sulphuric acid at 150° C.; from the reaction product after being rendered alkaline, para-chinanisol is distilled off, this on treatment with reducing-agents takes up four hydrogen atoms, forming the base thalline. This forms an oily liquid of strongly basic properties, uniting with acids, forming salts.

THALLINE SULPHATE forms a white, crystalline powder, soluble in 7 parts of cold water, 100 parts of alcohol and insoluble in ether. Oxidizing agents, as the halogens, argentic and mercuric nitrate, ferric chloride, etc., produce a bright green color. Internally thalline sulphate is an antipyretic in doses of 0.129 to 0.5 Gm. (2 to 8 grains); externally as an antiseptic injection (4 to 8 grains to the ounce).

THALLINE TARTRATE is a yellowish, crystalline powder, soluble in 10 parts of cold water, almost insoluble in alcohol and ether. Employed for like purposes as the sulphate.

THEOBROMIN. $C_7H_8N_4O_2$.

An alkaloid occurring in the seeds of *Theobroma cacao* (1.5%), obtained from the pressed cacao mass by mixing with slaked lime and exhausting with 8% boiling alcohol. It is a white crystalline powder, slightly soluble in water, alcohol and ether. Theobromin is a homologue of caffeine, differing in containing one CH_2 group less; it unites readily with alkalies forming soluble salts (see Diuretin). Because of its insolubility, theobromine is unsuitable for use, but is employed in form of a double salt. In physiological action it resembles caffeine, being, however, free from any irritating action on the nerve centers.

THEOBROMIN-LITHIUM-LITHIUM-SALICYLATE. See Uropherin.

THERMIFUGIN. $C_9H_7N(CH_3)(OH)COONa$.

Synonym : Methyl-trihydro-oxychinolin-carboxylate of sodium.

This compound forms colorless crystals, which are readily soluble in water, the solution becoming brown on standing. Employed as an antipyretic in doses of 0.1 to 0.25 Gm. (1.5 to 3.8 grains).

THERMIN. $C_{10}H_{11}.NH_2$.

Synonyms : Tetrahydro-b-naphthylamine.

Obtained by the action of metallic sodium on a solution of b naphthylamin in amyl alcohol. Thermin is a colorless liquid which with hydrochloric acid form colorless, soluble crystals, which melt at 237° C. (458.6° F.). Recommended by Filehne as a mydriatic; further, nothing definite is known concerning this substance.

THERMODIN. $CO.OC_2H_5.NCOH_4OCCH_3.OC_2H_5$.

Synonym : Acetyl-para-ethoxy-phenyl-urethane.

This derivative of urethane was introduced to replace neurodine, which is too powerful and rapid in its effects. Thermodin is a white crystalline powder, melts at 86° to 88° C. (186.8° to 190.4° F.), and is almost insoluble in cold water. Recommended as a mild antipyretic, free from any unpleasant effects ; given in doses of 0.32 to 1 Gm. (5 to 15 grains).

THERMOTAXINE.

A proprietary analgesic and antipyretic.

THILANIN.

This is a sulphurated lanolin, obtained by heating lanolin with sulphur at 230° C., and subsequently washing. It forms a brown unctuous mass, which contains about 3 per cent. of sulphur. Thilanin is employed as an application in various skin diseases.

THIOFORM. See Dithiosalicylic Acids.

THIOL.

A synthetic product of hydrocarbons obtained in a similar manner to ichthyol. The tarry oils obtained by the destructive distillation of peat are heated with sulphur at high temperature, the unsaturated hydrocarbons which unite with the sulphur are removed and by the action of sulphuric acid at a low temperature converted into sulphonated compounds called thiol, which is then purified by washing and dialysis and evaporated (in vacuo) to an extractive consistence (Thiolum Liquidum) or to dryness (Thiolum Siccum.

Thiol forms either a brownish-black, thick liquid (containing about 25 per cent. of dry residue) or a brownish-black powder, which is soluble in water and alcohol. It is precipitated from its aqueous solutions by mineral acids, metallic salts or alkali earths. Thiol is employed in the treatment of various skin diseases, its discoverers recommending it as a substitute for ichthyol. As an ointment the strength usually employed is 10 to 50 per cent. The dry thiol, which is about 2½ times the strength of the liquid, when mixed with starch is used as a dusting powder. Internal dose is 0.13 to 0.6 Gm. (2 to 10 grains.).

THIOLIN. See Thiolinic Acid.

THIOLINIC ACID.

Synonym : Thiolin.

This is prepared by boiling together linseed oil (6 p.) and sulphur (1 p.); the sulphurated linseed oil which is thereby formed is warmed with sulphuric acid until solution takes place, the oily product is poured into water and washed to remove the sulphuric and sulphurous acids.

Thiolinic acid forms a dark-green mass, of extractive consistency, and a peculiar mustard-like odor, insoluble in water, but soluble in alcohol.

The sodium salt, which constitutes a soluble powder, is preferred to the above.

The medicinal properties of thiolin are similar to those of thiol and ichthyol.

THIO-OXY-DIPHENYLAMINE. See Sulphaminol.

THIOPHENE DI-IODIDE. $C_4H_2I_2S$.

Obtained by the action of iodine and iodic acid on thiophene. Forms crystalline plates, insoluble in water, very soluble in chloroform, ether and warm alcohol, melting at 40.5° C. (104.9° F.); containing 75.5 per cent. of iodine and 9.5 per cent. of sulphur. Thiophene di-iodide is employed externally as a powder and gauze in all instances where iodoform might be applied.

THIORESORCIN. $C_6H_4(OS)_2$.

A sulphur derivative of resorcin, obtained by fusing one molecule of resorcin with two molecules of sulphur. A yellowish gray powder, insoluble in water; recommended as an iodoform substitute, but its use is followed by unpleasant symptoms.

THIOSALICYLIC ACID. $C_6H_4(SH)COOH$.

This is prepared from amido-benzoic acid, $(C_6H_4(NH_2)COOH$, by the action of nitrous acid and su phu:etted hydrogen. It is employed like salicylic and sulpho-salicylic acids as an antiseptic.

THIOSAPOL. $C_{18}H_{34}SO_2$.

A sulphuretted soap, prepared by heating unsaturated fats or fat acids such as oleic acid, with sulphur to a temperature of 120° to 160° C. Sulphur enters into combination, the product containing about 10 per cent. Soap containing sulphur in this state of combination is very serviceable in treatment of skin diseases.

THIOSINAMINE. $CS(NH_2)NH.C_3H_5$.

Synonyms: Allyl-sulpho-urea; Rhodallin; Allyl-sulpho-carbamide.

On heating a mixture of mustard oil (3p.), alcohol (3p.) and ammonia (6p.) at a temperature of 50° C., the pungent odor of the oil disappears, and on cooling crystals of thiosinamine are deposited. This forms colorless crystals of a slight alliaceous odor, melting at 74° C. (165.2° F.), very soluble in alcohol, water and ether.

Employed in treatment of lupus, in form of subcutaneous injections of 15 to 20% alcoholic solution.

THIURET. $C_8H_7N_3S_3$.

A sulphurated compound obtained by the oxidation of phenyl-dithio-biuret ($C_8H_9N_3S_2$). Forms a light, inodorous, crystalline powder, insoluble in water, quite soluble in alcohol and ether; in contact with alkalies (warmed) it gives up its sulphur. Thiuret, because of its kalyseptic and germicidal properties, is recommended as a substitute for iodoform. Various salts of thiuret have been prepared such as the *phenolsulphona'e, hydrochloride, hydrobromide, salicylate,* etc. These are more soluble in water than the base, and insoluble in ether and the oils. Their aqueous solutions give a violet coloration with ferric chloride and a white precipitate of the base (thiuret) on addition of **aqua ammoniæ**.

THYMACETIN. $\dfrac{CH_3}{C_3H_7} > C_6H_2 < {}^{OC_2H_5}_{NH(CH_3CO)}$.

By the action of nitric acid, thymol is converted into nitro-thymol, from which a sodium salt is prepared, this on heating with ethyl chloride under pressure yields nitro-thymol-ethyl-ether, which on reduction and acetification yields thymacetin. This forms a white crystalline powder, melting at 136° C. (276.8° F.), slightly soluble in water and freely in alcohol. Employed in treatment of neuralgia in doses of 0.19 to 0.64 Gm. (3 to 10 grains); it is said to produce unpleasant effects.

THYMENTHOL.

A proprietary antiseptic.

THYMOZONE.

A proprietary antiseptic.

THYRADEN. (Extractum Thyreodeæ-Haaf.)

This preparation is an extract of the **thyroid** gland, prepared according to an improved method by Dr. Haaf. It is said to possess all the active principles of the gland. Its non-toxic property and lack of odor are claimed as chief advantages over other similar preparations. It is so diluted with sugar of milk that one part of the extract is equivalent to two parts of the fresh glands.

THYROPROTEIN.

Notkine has recently isolated from the thyroid body an albuminoid which in properties and composition differs from all other albuminoids hitherto described. Under certain conditions this new body splits up, yielding a carbohydrate which is transformed with difficulty into a reducing substance. With a fairly strong solution of ferric chloride the albuminoid assumes a gelatinous consistence ; tannin precipitates it in the form of thick flakes or as a transparent gelatinous substance, according to the strength of the solution. It dissolves in weak acids, and is precipitated by alcohol, the precipitate rapidly becoming insoluble in water. The author has named this substance "thyroprotein." From experiments on animals it appears to be very toxic, and is slowly eliminated. It acts at first as an excitant, afterward as a paralyzant, probably on the central nervous system. The author regards this albuminoid as identical with the poison which accumulates in the organism after the extirpation of the thyroid body ; it is therefore not a secretion of the gland, but a toxalbumin which it is the function of the thyroid gland to neutralize by means of the peculiar ferment it elaborates, and so prevent the accumulation of the poison in the animal organism.

TOLYL-ANTIPYRIN. See Tolypyrin.

TOLYPYRIN. $C_6H_4CH_2N \begin{smallmatrix}CO.CH\\NCH_3.CCH_3\end{smallmatrix}$.

Synonyms : Tolylantipyrin; Beta-tolyl-dimethyl-pyrazolon.

Antipyrin is the phenyl (C_6H_5) derivative of di-methyl-pyrazolon ($N \begin{smallmatrix}COCH\\NCH_3.CCH_3\end{smallmatrix}$) while tolypyrine is the tolyl ($C_6H_4C_2H$) derivative of the same, or the latter may be considered as antipyrin in which one hydrogen atom of the phenyl radicle is replaced by a methyl group. This compound forms colorless crystals, melting at 136° to 137° C. (276.8° to 278.6° F.), soluble in 10 parts of water, readily in alcohol and insoluble in ether. Tolypyrin gives the same color reactions with ferric chloride and nitrous acid as antipyrine, and, like the latter, is employed as an antipyretic, antirheumatic and antineuralgic in the same doses, 0.5 to 2 Gm. (8 to 30 grains).

TOLYPYRIN SALICYLATE. See Tolysal.

TOLYSAL. $C_6H_4CH_2N \begin{smallmatrix}COCH\\NCH_2.CCH_3.C_7H_6O_2\end{smallmatrix}$.

Synonym : Tolypyrin Salicylate.

Obtained by fusing together equimolecular quantities of tolypyrin and salicylic acid and crystallizing from alcohol. Forms colorless crystals, melting at 101° to 102° C. (213.8° to 215.6° F.), almost insoluble in water and readily soluble in alcohol.

Employed in chronic and **acute** rheumatism and rheumatic neuralgia in doses of 1 to 2 Gm. (15 to 30 grains.).

TRAUMATOL. C_7H_7IO.

An iodocresol, of purple-red color, obtained by the action of iodine on cresol. It is expected to find use as a substitute for iodoform, being odorless, non-toxic, antiseptic and non-irritating.

TREFUSIA.

A dark red-brown, soluble, granular powder, obtained by drying **defibrinated blood.** Employed as a natural iron albuminate in chlorosis.

TRI-BROM-ALDEHYDE HYDRATE. See Bromal Hydrate.

TRI-BROM-ANILIN HYDROBROMIDE. See Bromamide.

TRI-BROM-PHENOL. See Bromol.

TRI-BROM-SALOL. See under Salol.

TRI-CHLOR-ACETIC ACID. CCl_3COOH.

This is obtained by the action of chlorine on glacial acetic acid, or by the oxidation of anhydrous chloral by means of fuming nitric acid.

Trichloracetic acid occurs in colorless, rhombic crystals, very hygroscopic, of a slightly penetrating odor ; melting at 55° C. (131° F.). Very soluble in water and alcohol.

It is employed as a caustic in 10 to 50 per cent. solution, also as a **test reagent for albumin in urine.**[*]

TRICHLORPHENOL. See under Chlorphenol.

TRICRESOL. See under Cresol.

[*] COBLENTZ's Handbook of Pharmacy, p. 474.

TRICRESOLAMINE.

A mixture of equal parts of ethylene-diamine and tricresol. Used as an antiseptic.

TRIFORMAL. See Formalin.

TRI-IODO-META-CRESOL. See Losophan.

TRI-METHYL-ETHYLENE. See Pental.

TRIONAL. $C_2H_5.CH_3.C.(SO_2C_2H_5)_2$.

Synonym: Di-ethyl-sulphon-methyl-ethyl-methane.

This is an analogue of sulphonal and differs in the substitution of a methyl by an ethyl group (see Tetronal). The method of preparation is the same as that of sulphonal, only that methyl-ethyl-ketone (CH_2-CO-C_2H_5) is employed in place of acetone. Trional forms colorless, shining, tabular crystals, melting at 76° C. (168.8° F.), soluble in 320 parts of cold and freely in hot water, very soluble in alcohol and ether. This compound is a more powerful hypnotic than sulphonal and is less liable to produce ill effects. It is preferred to tetronal as a reliable and safe hypnotic. Dose, 1 to 2 Gm. (15 to 30 grains).

TRIPHENIN. $C_6H_4C_2H_5O.NH(CH_3.CH_3CO)$.

This is a homologue of phenacetin, forming an insoluble powder, melting at 120° C. (248° F.). It is recommended as an antipyretic and antineuralgic in doses of 0.3 to 0.6 Gm. (4.5 to 10 grains).

TROPA-COCAINE HYDROCHLORIDE. $C_8H_{14}NO.(C_6H_5CO)HCl$.

Synonym: Benzoyl-pseudotropein hydrochloride.

This alkaloid occurs with cocaine and other bases in the small Java coca leaves; prepared synthetically by Liebermann. Forms white needles, melting at 271° C. (519.8° F.), and is readily soluble in water. Tropacocaine in 2 or 3 per cent. solutions produces more rapid anæsthesia, is less toxic, and more reliable than cocaine (Ferdinando and Chadbourne).

TUMENOL.

Synonym: Sulphotumenolic Acid.

This compound, which is closely allied to ichthyol, is obtained by treating (sulphonating) the unsaturated hydrocarbons of mineral oils with sulphuric acid, the resulting product is dissolved in water and separated in pure form by the addition of salt. The tumenol-sulphonic acid thus obtained is known as "Commercial Tumenol," being a dark-brown, almost black, acid fluid; this on being neutralized with soda and extracted with ether, yields the "Tumenol-Sulphone" (Tumenol Oil), which is a thick, dark yellow, syrupy fluid, with a bitter taste and insoluble in water. This latter is prepared in the powder form, known as "Tumenol-Sulphonic Acid," being of dark color, inodorous, slightly bitter, and readily soluble in water. Tumenol is employed in treatment of all forms of pruritis and also eczema, either as a 5 to 10 per cent. solution (ether-alcohol, water or glycerin) ointment, paste or dusting-powder. The tumenol oil is frequently painted directly over the diseased surfaces.

SODIUM TUMENOL-SULPHONATE is a combination of sulphotumenolic acid and sodium. A dark-colored, dry powder, soluble in water, and applied in all instances above cited.

TUSSOL.

Synonym: Antipyrine **Mandelate** or Phenyl-glycolate.

This new salt of antipyrine is recommended as being superior to antipyrine itself in the treatment of whooping cough. Given in doses of 0.05 to 0.1 Gm. (7-10 to 1½ grains) for children under one year of age; 0.1 Gm. (1½ grains) for 1 to 2 years; 0.25 to 0.4 Gm. (3.8 to 6 grains) for 2 to 4 years, and 0.5 Gm. (8 grains) for 5 years and above.

ULYPTOL.

Synonym: Eulyptol.

This is a name given to a mixture of phenol 1 part, salicylic acid 6 parts, and eucalyptus oil 1 part.

URALINE. See Uralium.

URALIUM. $CCl_3CH:OH.NHCO_2C_2H_5$.

Synonyms: Chloral Urethane; Uraline.

To a solution of urethane (q. v.) in melted chloral hydrate, concentrated hydrochloric acid is added; after 24 hours it congeals to a solid mass, which is then washed with sulphuric acid, followed by water, leaving an oil, which, on standing, crystallizes

Uralium constitutes a white powder, melting at 103° C. (217.4° F.), insoluble in cold water, very soluble in alcohol and ether; when boiled with water it decomposes into chloral and urethane. Recommended as a hypnotic in doses of 2 to 3 Gm. (30 to 45 grains).

URETHANE. $CO < ^{NH_2}_{OC_2H_5}$.

Synonym: Ethyl-Urethane ; Ethyl Carbamate.

This compound, an ethyl ether of carbamic acid $\left(CO < ^{NH_2}_{OH} \right)$ is obtained by heating a salt of urea with ethyl alcohol under pressure at a temperature of 120° to 130° C. Forms colorless, odorless prisms or scales, melting at 50° to 51° C. (122° to 123.8° F.) soluble in 1 part of water, 0.6 part of alcohol, 1 part of ether and 1.5 parts of chloroform. Urethane is an excellent hypnotic, being free from by or after effects ; dose is 1 to 2 Gm. (15 to 30 grains.)

UROPHERIN. $C_7H_7N_4O_2Li + C_6H_4(OH)COOLi$.

Synonyms: Lithium Diuretin ; Theobromine-lithium-lithium-Salicylate.

This double salt is analogous to diuretin, differing only in the substitution of lithium for sodium. It is prepared by rubbing together theobromine with lithium hydroxide and an equivalent quantity of lithium salicylate, with sufficient water, and then dried. It is a white powder soluble in 5 parts of water. Employed as a diuretic in doses of 1 Gm. (15 grains).

UTROPIN. $(CH_2)_6N_4$.

Synonym: Hexa-methylene-tetramin.

A compound produced by the action of formaldehyde on ammonia. Urotropin increases the excretion of the urine and of uric acid, the solution of the urates beginning within 24 hours of the ingestion of the medicament. It may be given in doses of 6 Gm. (90 grains) daily to adults, the single dose being from 1 to 1.5 Gm. (15 to 23 grains).

VALERYL-PHENETIDINE. See under Phenetidine.

VANILLIN. $C_6H_3OH.OCH_3.CHO$.

Synonym: Methyl-protocatechuic aldehyde.

This odorous principle, which is found in the vanilla pods, also occurs in small quantities in gum benzoin, asparagus, raw beet sugar and the wood of many plants, synthetically prepared from coniferin, a glucoside, and also from eugenol. Vanillin occurs in acicular crystals, melting at 80° to 81° C (176 to 177.8° F.), soluble in alcohol, glycerin, ether and chloroform, only slightly soluble in water. It possesses the odor and taste of vanillin. Employed chiefly as an odoriferous and flavoring agent.

VASELON.

Solution of stearin and margarin in neutral mineral oil. Ointment base.

VASOGEN.

A new ointment basis which has been introduced in Germany. It is said to be an oxygenated vaseline (vaseline with free oxygen), but another statement is that it contains about 25 per cent. of olein, saponified with anhydrous ammonia and mixed with vaseline, and brought to a suitable consistency with vaseline oil.

VICOSIN.

A mixture of caramel and extract of saponaria, used for producing a permanent foam on beer.

VIERIN.

An amorphous, white, bitter principle, of aromatic odor, obtained from the bark of *Rentzia vellozii*. It melts at 120° C. (248° F.), and is readily soluble in alcohol and chloroform. Employed as a quinine substitute in doses of 0.1 to 0.2 Gm. (1½ to 3 grains).

VUTRIN.

A concentrated meat extract in powder form, one part of which represents the nutritive value of four parts of beef.

XYLENOL-SALOLS. $C_6H_4(OH)COO\ C_6H_3(CH_3)_2$.

By the action of dehydrating agents upon a mixture of equal molecules of salicylic acid and xylenol (o-m-or-p.), ortho, meta or para-xylenol salicylates are formed. These are insoluble in water and soluble in alcohol; employed like salol as intestinal disinfectants.

ZAPON LAC.

A new quick-drying lac or varnish which consists of gun cotton dissolved in a mixture of amyl acetate and amylic alcohol. It is coming into use in pharmacy abroad as a varnish for unguent boxes, etc., especially the new and elegant celluloid boxes.

ZINC COMPOUNDS.

BORATE. $ZnB_4O_7+7H_2O$. Prepared by interaction between hot solutions of 5 parts of zinc sulphate in 50 parts of water and 4 parts of borax in 100 parts of water. An amorphous, white powder, which is employed as an antiseptic dusting powder for wounds.

CHRYSOPHANATE. Forms a brownish-red powder, which is readily soluble in water which has been rendered slightly alkaline. Recommended as antiseptic dusting powder.

GYNOCARDATE. A yellow, granular powder, insoluble in water and dilute acids, readily soluble in alcohol and ether. Recommended in treatment of psoriasis, prurigo and other skin diseases.

HÆMOL. See under Hæmol.

IODATE. See Iodic Acid.

MERCURIC-CYANIDE. $Zn_4Hg(Cu)_{10}$. A white insoluble powder recommended as a non-irritating antiseptic.

PERMANGANATE. Occurs in crystals closely resembling those of the potassium salt; hygroscopic and soluble in water. This salt is employed in all instances where zinc sulphate is indicated; its solutions being of the strength 0.05 Gm. to 200 Cc. of water (7-10 grain to 6.8 fld. ozs). Care should be taken not to triturate this salt with organic substances or dispense it in solutions containing alcohol or organic extracts (See Coblentz's Handbook of Pharmacy, pp. 392-396).

SALICYLATE. $(C_7H_5O_3)_2Zn+H_2O$. Sodium salicylate 34 parts and zinc sulphate 29 parts are boiled for a short time with 125 parts of water; on cooling a solid crystalline mass separates, which, after washing with a little cold water, is recrystallized. Forms colorless crystals which are soluble in 25.2 parts of cold and readily in boiling water, soluble in 36 parts of ether and 3.5 parts of alcohol. Recommended as antiseptic dusting powder and wash in various skin diseases.

SOZOIODOL. $(C_6H_2I_2(OH)SO_3)_2Zn+6H_2O$. See under Sozoiodol.

SUBGALLATE is an odorless, non-toxic, non-irritant, greenish-gray, neutral powder, insoluble in the ordinary solvents, containing 44 per cent. of zinc oxide, and 56 per cent. of gallic acid. Used internally and externally. Internally it is recommended in doses of ½ to 4 grains (0.03 to 0.25 Gm.) in fermentive disorders of the intestines and in night sweets. Externally, it has been used in eczema, fresh and septic wounds, otorrhea, gonorrhea and hemorrhoids. It is applied pure or diluted with indifferent powders or ointments. As an injection it is suspended in water and mucilage, 1 to 16.

SULPHOCARBOLATE. $C_6H_4(OH)SO_3)_2Zn+8H_2O$. By the action of concentrated sulphuric acid on phenol at 90° C., para-phenol-sulphonic acid is formed; this is neutralized with barium carbonate, and the resulting barium sulphocarbolate on being brought into reaction with an equivalent amount of zinc sulphate in solution yields zinc sulphocarbolate and the insoluble barium sulphate. The filtrate is evaporated and crystallized. This salt forms colorless, rhombic prisms or scales, soluble in water and alcohol. Employed as an antiseptic wash in all instances where zinc sulphate or carbolic acid is indicated.

SULPHYDRATE. $Zn(SH)_2$. A white precipitate, which should be kept under water, since it readily decomposes on becoming dry. Recommended by Barduzzi externally and internally in the treatment of chronic eczema, psoriasis and vegeto-parasitic dermatoses. Internally the dose is 0.03 to 0.13 Gm. (½ to 2 grains), externally in ointment form (10%) combined with lanolin and lard (2:3).

ZYMOIDIN.

Said to be composed of the oxides of zinc, bismuth, and aluminum with iodine, boric acid, carbolic acid, gallic acid, salicylic acid, quinine, etc. Recommended as an antiseptic in the form of dusting powder, ointment, solution or bougie.

THE NEWER REMEDIES

PIPERAZIN
SCHERING.

The characteristic property of this compound of dissolving uric acid and urate concretions, forming freely soluble urates, accounts for the beneficial results and consequent extended use in the treatment of uric acid diathesis, *gout,* arthritis and renal colic. Dose, 1 gramme per day, dissolved in a quart of carbonated water, which patient may drink ad libitum.

PIPERAZIN
WATER
SCHERING.

For the convenience of physicians we supply *Schering's Piperazin Water,* a mildly carbonized water containing one gramme each of Piperazin and Phenocoll, yielding a preparation of double effect: *First,* the pain-relieving action of phenocoll, and *second,* the uric acid solvent (radical curative) action of Piperazin. The dose per day is the contents of *one bottle* of *Piperazin Water,* the best effect being obtained by drinking it in wineglassfuls at frequent intervals during the day. The water should be kept at a moderately warm temperature, as the system will then assimilate it more readily.

SALIPYRIN
RIEDEL.

Is the salicylic acid salt of Antipyrin in a stable form. It acts promptly and never fails in Influenza and Rheumatic Affections. It has been proven to be free from the untoward side effects often exhibited by Antipyrin and exercises a quieting influence on the nervous system.

THIOL
RIEDEL.

Is a synthetical product containing 12 per cent of sulphur. It is chemically identical with Ichthyol and therapeutically superior because pure, uniform, odorless and non-toxic. It is free from the unpleasant odor of Ichthyol and does not stain linen; it is also free from the irritant effect which Ichthyol often manifests. Liquid Thiol is the staple form for regular use in salves, ointments, plasters, etc., for application in cutaneous affections. It is indicated for use as Alterative, Anodyne, Astringent, Resolvent, Antiphlogistic, Gastric and Renal Tonic, etc.

LYSOLUM
PURUM

Is non toxic, it makes a clear solution in water and is superior to Carbolic Acid and Creolin as a bacter icide; it is cheaper than Carbolic Acid because a 2 per cent solution is equal to a 5 per cent solution of Carbolic Acid, crystals

All these facts, borne out by the liberal contributions on the subject to current medical literature, have made Lysol the favorite antiseptic and disinfectant in obstetrical and surgical practice.

A full descriptive pamphlet containing clinical reports and extracts on Lysol sent to physicians free on request.

LEHN & FINK, -- NEW YORK.

ALBUMINATE OF IRON

Dr. DREES'

FOR ANAEMIA AND CHLOROSIS

Has gained the confidence of the medical profession. It has been in ever increasing demand since its introduction 10 years ago, because it is *positively efficient* and is readily borne by the digestive organs.

The dose for adults is one drachm three times a day, for children 10 to 30 drops generally taken with milk.

PAPAIN

L & F.

Is highly esteemed as a digestive aid ; unlike Pepsin or Pancreatin, it will dissolve albumen and fibrine in *neutral* and *alkaline* media, and hence is active in *both stomach* and *intestines*. Dose, 3 to 5 grains. *Papain, L. & F.*, has yielded excellent results as a *diphtheritic membrane solvent* in 5 per cent solution.

Pamphlets on Papain, the Vegetable Pepsin, Its Origin, Properties and Uses, also Samples, mailed free on request.

PLASTER-MULLS

Dr. UNNA'S

FOR DERMATOLOGICAL USE.

Prepared according to directions of the originator, Dr. P. G. Unna, the eminent dermatologist by P. Beiersdorf & Co., Hamburg, Germany.

Plaster Mulls are an ideal improvement on the ordinary spread plasters. The medicament is dissolved in a minimum quantity of a suitable vehicle, and is evenly spread on the thinnest possible layer of gutta-percha, backed by fine gauze (or mull). The gutta-percha layer hermetically seals the covered skin surface and prevents transpiration (or exhalation) of the pores; thereby the medicament is more deeply absorbed, and the therapeutic effect is more intense and powerful. The stratum of plaster is very thin, consisting of a non irritant vehicle and the medicament; the former in smallest possible proportion, and the latter in exact dosage perfectly dissolved, so that every particle will act its full remedial part in every particular; these Plaster-Mulls are vastly superior to the old style thickly spread plasters and ointments.

Unna's Plaster-Mulls have been used and favorably endorsed by many authorities, such as: Nega, Will, Janowsky, McCall, Anderson, Frank, Morrison, J. C. McGuire, Schwimmer, Chotzen, Bulkley, Borck, and many others.

We supply Plaster Mulls in 1 metre rolls, in tin canisters. Price list mailed on application

STREPTOCOCCUS ANTITOXIN

GIBIER S

Obtained from horses immunized against Streptococcus Erysipelatis. It is effectively employed in Septi. caemia, Puerperal **Fever** and Erysipelas. Put up in vials containing 25 ccm.

LEHN & FINK, — NEW YORK.

New York Quinine

AND

Chemical Works,

(LIMITED).

QUININE	ALOIN
COCAINE	CODEINE
MORPHINE	ACETANILID.

AND A GENERAL ASSORTMENT OF MEDICINAL CHEMICALS.

N
Y
Q
Our products are unsurpassed in quality and appearance, they are carried in stock by wholesale druggists generally, and your preference is respectfully solicited.

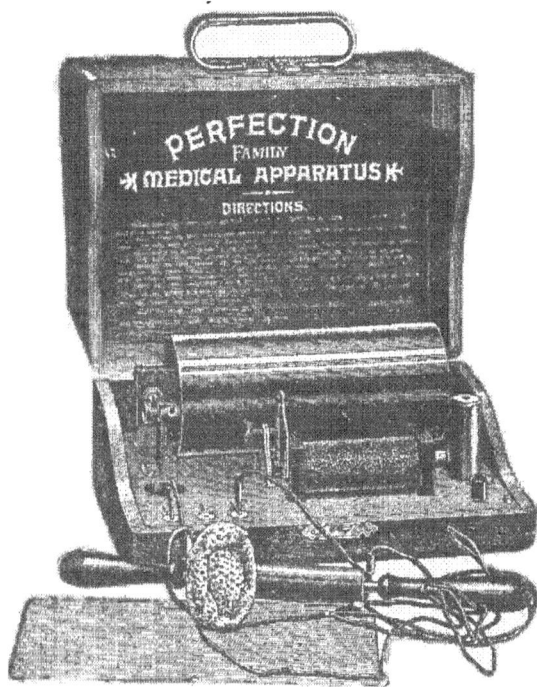

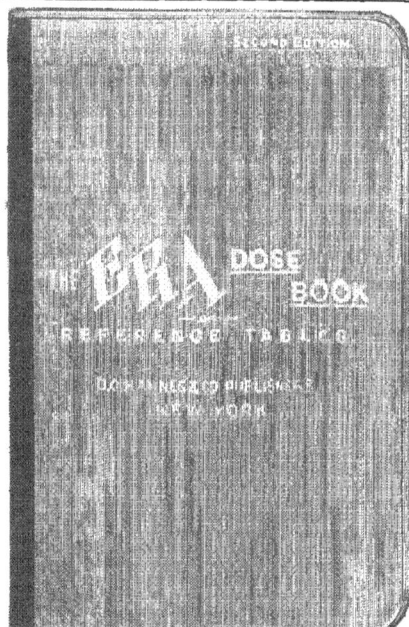

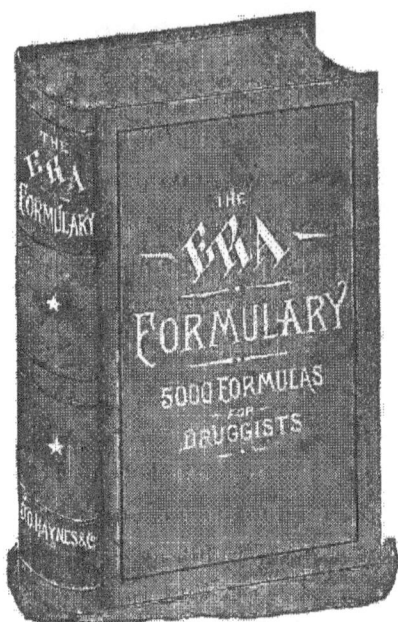

A FULL LINE

OF

www.ingramcontent.com/pod-product-compliance
Lightning Source LLC
Chambersburg PA
CBHW021948190326
41519CB00009B/1182